My Book

This book belongs to

Name: _____

©All rights reserved-Math-Knots LLC., VA-USA   www.math-knots.com

Copy right © 2019 MATH-KNOTS LLC

All rights reserved, no part of this publication may be reproduced, stored in any system or transmitted in any form, or by any means, electronic, mechanical, photocopying, recording, or otherwise without the written permission of MATH-KNOTS LLC.

Cover Design by :
Gowri Vemuri

First Edition :
September, 2020

Author :
Gowri Vemuri

Editor :
Ritvik Pothapragada

Questions: mathknots.help@gmail.com

NOTE : TJHSST (Thomas Jefferson High School for Science and Technology) or VDOE (Virginia Department of Education) is neither affiliated nor sponsors or endorses this product.

Dedication

This book is dedicated to:

My Mom, who is my best critic, guide and supporter.

To what I am today, and what I am going to become tomorrow,

is all because of your blessings, unconditional affection and support.

This book is dedicated to the

strongest women of my life,

my dearest mom

and

to all those moms in this universe.

G.V.

QUANT - Q

PREFACE

Thomas Jefferson High School for Science and Technology (TJHSST) admission test is based on the below rubric

Test	Math- Quant Q	Reading- Aspire	Science- Aspire
Time	50 minutes	65 minutes	60 minutes
Number of Questions	28	32	40
Measures	<ul><li>Pattern Recognition</li><li>Probability Combinatorics</li><li>Out-of-the Box Algebra</li><li>Geometry and Optimization</li></ul>	<ul><li>Key Ideas and Details</li><li>Craft and Structure</li><li>Integration of Knowledge and Ideas</li></ul>	<ul><li>Interpretation of Data</li><li>Scientific Investigation</li><li>Evaluation of Models</li><li>Inferences</li><li>Experimental Design</li></ul>

   This book (Book 1 Volume 1) helps the students to practice on Pattern recognition, Out of Box algebra and Geometry. Few advanced topics are added as well. Each topic is subdivided into smaller topics with 15 practice questions in each. A total of 1000+ questions to practice for the students.

Book 1 Volume 2 will cover the remaining topics Probability, Combinatorics, Geometry and Optimization.

   Notes on various topics is provided in the beginning of the book and Answer keys in the end.

INDEX

Preface	1 - 12
Notes	13 - 32
Formulae	33 - 38
Prime Factors	39 - 41
Numerical expressions	42
Simplify expressions	43
Evaluate expression	44
Exponential expressions	45
Distance between 2 points	46
Midpoint	47 - 50
Equation of a straight line	51
Slope 2 points	52 - 54
Slope intercept form	55 - 57
Slope graph	58 - 59
Find the slope	60
Parallel line slope	61
Perpendicular line slope	62
Radicals 1	63 - 65

INDEX

Radicals 2	66 - 68
Inequalities	69
One step word problems	70 - 71
Circle area	72 - 73
Volume Sphere	74 - 75
Volume rectangle square prisms	76 - 77
Volume Cone Cylinder	78 - 79
Missing angle 1	80 - 81
Missing angle 2	82 - 83
Reflection	84 - 85
Rotation	86 - 87
Translation	88 - 90
Warm up 8	91 - 94
Practice test 8	95 - 100
Warm up 9	101 - 102
Practice test 9	103 - 108
Warm up 10	109 - 110
Practice test 10	111 - 116

QUANT - Q

INDEX

Warm up 11	117 - 118
Practice test 11	119 - 124
Warm up 12	125 - 126
Practice test 12	127 - 132
Warm up 13	133 - 134
Practice test 13	135 - 140
Warm up 14	141 - 142
Practice test 14	143 - 148
Warm up 15	149 - 150
Practice test 15	151 - 156
Answer Keys	157 - 172

NUMBER SYSTEM

Real Numbers:

A number that can be represented on a number line is called a real number.

Rational Numbers (Q):

The numbers can be expressed in the form of $\frac{p}{q}$ (p, q ∈ Z, q ≠ 0) are called Rational numbers.

Irrational Numbers:

The numbers which are not rational numbers are called irrational numbers. i.e., The numbers that cannot be expressed in the form $\frac{p}{q}$ where q is a non zero integer are called irrational numbers.

Example: π

Integers: I or Z = { ……. –3, –2, –1, 0, 1, 2, 3,……}

Whole Numbers: W = {0, 1, 2, 3, 4, 5, 6, …….}

Natural Numbers: N = {1, 2, 3, 4, 5, 6,…….}

Divisibility rules

A number is said to be "divisible" by another if the second number divides evenly into the first. Example, the number 10 is divisible by number 1,2 and 5 evenly into 10.

Divisibility rules #2

1. All even numbers are divisible by 2. Example: Any number ending in 0,2,4, 6 or 8.

Divisibility rules #3

1. Add all the digits in the number.

2. If the sum of the digits is divisible by 3, the number is divisible by 3.

3. Example : 13461 : 1 + 3 + 4 + 6 + 1 = 15. 15 is divisible by 3.
   so , 13461 is divisible by 3

Divisibility rules #4

1. Check the last two digits of the given number is divisible by 4 ?

2. If yes, the number is divisible by 4

3. Example : 34336 ends in 36 which is divisible by 4.
   so , 34336 is divisible by 4.

Divisibility rules #5

1. Numbers ending in 5 or 0 are always divisible by 5.

Divisibility rules #6

1. If a number is divisible by 2 and 3, then it is divisible by 6.

Divisibility rules #9

1. Add all the digits in the number.

2. If the sum of the digits is divisible by 9, the number is divisible by 9.

3. Example : 31905 : $3 + 1 + 9 + 0 + 5 = 18$. 18 is divisible by 9.
   so, 31905 is divisible by 9.

Divisibility rules #10

1. Any number ending with a digit 0 is divisible by 10.

EXPONENTS PROPERTIES

(1) The square of a real number is always positive.
If $m, n \in Z; a, b \in R \ (a \neq 0)$

(i) $a^{-m} = \dfrac{1}{a^m}$ (ii) $a^m \cdot a^n = a^{m+n}$

(iii) $\dfrac{a^m}{a^n} = a^{m-n}$ (iv) $(a^m)^n = a^{mn}$

(v) $(ab)^m = a^m b^m$ (vi) $\left(\dfrac{a}{b}\right)^m = \dfrac{a^m}{b^m}$ $(b \neq 0)$.

(vii) $\sqrt[n]{a} = a^{1/n}$ (viii) $\sqrt[n]{a^m} = a^{m/n}$

(ix) $\sqrt[n]{\dfrac{a}{b}} = \dfrac{\sqrt[n]{a}}{\sqrt[n]{a}}$ (x) $\sqrt[m]{\sqrt[n]{a}} = a^{\frac{1}{mn}}$

(xi) $a^0 = 1 \ (a \neq 0)$

Example:

Simplify (i) $(27)^{5/3}$.
$(27)^{5/3} = (3^3)^{5/3} = 3^5 = 243.$ $(\because (a^m)^n = a^{mn})$

(ii) $\left(\dfrac{625}{16}\right)^{-5/4} = \left(\dfrac{625}{16}\right)^{-5/4} = \left(\dfrac{16}{625}\right)^{5/4}$ $(\because a^{-m} = \dfrac{1}{a^m})$

$= \left(\dfrac{2^4}{5^4}\right)^{5/4} = \left(\dfrac{2^5}{5^5}\right)$ $\left(\because \left(\dfrac{a}{b}\right)^m = \dfrac{a^m}{b^m}\right) = \left(\dfrac{2}{5}\right)^5$.

InEqualities and Absolute value of a Number:

If x is a real number, then its absolute value or modulus value is denoted by $|x|$ (read as mod x) and defined as follows,

$|x| = x$ when $x > 0$.

$|x| = -x$ when $x < 0$.

$|x| = 0$ when $x = 0$.

For real numbers A and B , an equation as $|A| = B$ when $B = 0$
has 2 solutions $A = B$ or $A = - B$.
If $B < 0$ then $|A| = B$ has no solution.

Set	Notation	Interval Notation
All real numbers between a and b, not including a or b	$\{x|\ a < x < b\}$	(a , b)
All real numbers greater than a, not including a	$\{x|\ x > a\}$	(a , ∞)
All real numbers less than b, not including b	$\{x|\ x < b\}$	(-∞ , b)
All real numbers greater than a, including a	$\{x|\ x \geq a\}$	[a , ∞)
All real numbers less than b, including b	$\{x|\ x \leq b\}$	(-∞ , b]
All real numbers between a and b, including a	$\{x|\ a \leq x < b\}$	[a , b)
All real numbers between a and b, including b	$\{x|\ a < x \leq b\}$	(a, b]
All real numbers between a and b, including a and b	$\{x|\ a \leq x \leq b\}$	[a , b]
All real numbers less than a or greater than b	$\{x|\ x < a$ and $x > b\}$	(-∞ , a) U (b , ∞)
All real numbers	$\{x|\ x$ is all real numbers$\}$	(-∞ , ∞)

Properties of inequalities :

1. Addition property
   If a < b, then a + c < b + c

2. Multiplication property
   If a < b and c > 0, then ac < bc
   If a < b and c < 0, then ac > bc

Example:

The value of $|7| = 7$ (7 > 0)

The value of $|-5| = -(-5) = 5$ (-5 < 0)

The value of $|x|$ is always positive.

Range of |x| is positive real numbers including zero.

Properties:

(i) If $|x| = a$ then $x = \pm a$.

(ii) If $|x| \leq a$ then $-a \leq x \leq a$.

(iii) If $|x| \geq a$ then $x \leq -a$ or $x \geq a$.

Example 1: If $|x| = 8$, then find the values of x.

Solution: If $|x| = a$; $x = \pm a$
$|x| = 8$, $x = \pm 8$.

Example 2: If $|x| \leq 9$, then find the range of x.

Solution: If $|x| \leq 9$, then $-a \leq x \leq a$
$|x| \leq 9 \Rightarrow -9 \leq x \leq 9$.

Example 3: If $|x| \geq 15$, then find the range of x.

Solution: If $|x| \geq a$, the $x \leq -a$ or $x \geq a$.

$|x| \geq 15 \Rightarrow x \leq -15$ or $x \geq 15$.

Example 4: If $|x+5| = 9$, find the value of x.

Solution: $|x| = a$, then $x = \pm a$

$|x+5| = 9 \Rightarrow x + 5 = \pm 9$.

$x + 5 = 9$ or $x + 5 = -9$

$x = 4$ or $x = -14$.

Example 5: $|3x - 7| = 22$, find the value of x.

Solution: $|3x - 7| = 22 \Rightarrow 3x - 7 = \pm 22$

$3x - 7 = 22$ or $3x - 7 = -22$

$3x = 29$ or $3x = -22 + 7$

$x = \dfrac{29}{3}$ or $3x = -15 \Rightarrow x = -5$

$x = \dfrac{29}{3}$ or $x = -5$.

Example 6: Solve $|x - 5| < 3$.

Solution: $|x| < a \Rightarrow -a < x < a$

$|x - 5| < 3 \Rightarrow -3 < x - 5 < 3$.

$\Rightarrow -3 + 5 < x < 3 + 5$

$2 < x < 8$.

Example 7: Solve $|3x - 10| > 43$.

Solution: $|x| > a \Rightarrow x < -a$ or $x > a$.

$|3x - 10| > 43 \Rightarrow 3x - 10 < -43$ or $3x - 10 > 43$.

$3x < -43 + 10$ or $3x > 43 + 10$

$3x < -33$ or $x > \dfrac{53}{3}$.

$x < -11$ or $x > \dfrac{53}{3}$.

Graph of $|x|$:

Since x is any real number, $|x|$ always positive, so graph of $|x|$ belongs to first and second quadrants only.

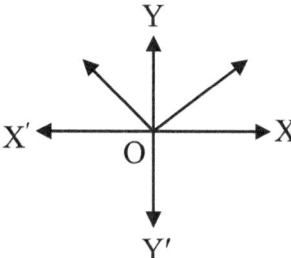

Solving radical equations
An equation containing terms with a variable in the radicand is called a radical equation.

Steps to solve the radical equation :
1. Isolate the radical expression on one side of the equal sign.
2. Move all other terms to other side of the equation.
3. If the radical is a square root , then square both sides of the equation.
   If the radical is a cube root , then rise both sides of the equation to the third power.
   If the radical is a nth root , then rise both sides of the equation to the nth power.
4. If the radical still exists repeat steps 1,2,3. Repeat the process until the radical is eliminated. .
5. Solve the equation.
6. Substitute the solution back for verification.

Solving absolute value equations
The absolute value of x is written as $|x|$. It has the following properties:
If $x \geq 0$, then $|x| = x$.
If $x < 0$, then $|x| = -x$.
For real numbers A and B, an equation of the form $|A| = B$, with $B \geq 0$, will have solutions when
A = B or A = −B.

If B < 0, the equation $|A| = B$ has no solution.
An absolute value equation in the form $|ax + b| = c$ has the following properties:
If c < 0, $|ax + b| = c$ has no solution.
If c = 0, $|ax + b| = c$ has one solution.
If c > 0, $|ax + b| = c$ has two solutions.

1. Isolate the absolute value expression on one side of the equal sign.
2. If $c > 0$, write and solve two equations: $ax+b=c$ and $ax+b=-c$.

The Discriminant

The quadratic formula not only gives the solutions to a quadratic equation $ax^2 + bx + c = 0$, but also describes the nature of the solutions. The discriminant, $b^2 - 4ac$ determines whether the solutions are real numbers or complex numbers. Below table describes the properties of the discriminant to the solutions of a quadratic equation.

Value of Discriminant	Results
$b^2 - 4ac = 0$	One rational solution (double solution)
$b^2 - 4ac > 0$, perfect square	Two rational solutions
$b^2 - 4ac > 0$, not a perfect square	Two irrational solutions
$b^2 - 4ac < 0$	Two complex solutions

polynomial equations

A polynomial of degree *n* is an expression of the type

$$a_n x^n + a_{n-1} x^{n-1} + \cdots + a_2 x^2 + a_1 x + a_0$$

where *n* is a positive integer and $a_n, \ldots, a_0$ are real numbers and $a_n \neq 0$.

Setting the polynomial equal to zero gives a **polynomial equation**. The total number of solutions (real and complex) to a polynomial equation is equal to the highest exponent *n*.

Solving Radical Equations:

An equation containing terms with a variable in the radicand is called a **radical equation**.

Given a radical equation, solve it.

1. Isolate the radical expression on one side of the equal sign. Put all remaining terms on the other side.

2. If the radical is a square root, then square both sides of the equation. If It is a cube root, then raise both sides of the equation to the third power. In other words, for an *n* the root radical, raise both sides to the *n*th power. Doing so eliminates the radical symbol.

3. Solve the remaining equation.

4. If a radical term still remains, repeat steps 1–2.

5. Confirm solutions by substituting the min to the original equation.

Notes

Set Indicated	Set-Builder Notation	Interval Notation
All real numbers between a and b, but not including a or b	$\{x \mid a < x < b\}$	(a, b)
All real numbers greater than a, but not including a	$\{x \mid x > a\}$	(a, ∞)
All real numbers less than b, but not including b	$\{x \mid x < b\}$	$(-\infty, b)$
All real numbers greater than a, including a	$\{x \mid x \geq a\}$	$[a, \infty)$

Set Indicated	Set-Builder Notation	Interval Notation
All real numbers less than b, including b	$\{x \mid x \leq b\}$	$(-\infty, b]$
All real numbers between a and b, including a	$\{x \mid a \leq x < b\}$	$[a, b)$
All real numbers between a and b, including b	$\{x \mid a < x \leq b\}$	$(a, b]$
All real numbers between a and b, including a and b	$\{x \mid a \leq x \leq b\}$	$[a, b]$
All real numbers less than a or greater than b	$\{x \mid x < a \text{ and } x > b\}$	$(-\infty, a) \cup (b, \infty)$
All real numbers	$\{x \mid x \text{ is all real numbers}\}$	$(-\infty, \infty)$

---

properties of inequalities

Addition Property — If $a < b$, then $a + c < b + c$.

Multiplication Property — If $a < b$ and $c > 0$, then $ac < bc$.

If $a < b$ and $c < 0$, then $ac > bc$.

These properties also apply to $a \leq b$, $a > b$, and $a \geq b$.

Number series

Sequences of numbers which follow specific patterns are called progression. Depending on the pattern, the progressions are classified as follows.

(i) Arithmetic Progression
(ii) Geometric Progression and
(iii) Harmonic Progression

ARITHMETIC PROGRESSION (A.P.)

Numbers (or terms) are said to be in arithmetic progression when each one, except the first, is obtained by adding a constant to the previous number (or term).

An arithmetic progression can be represented by $a, a + d, a + 2d, \ldots, [a + (n-1)d]$. Here, d is added to any term to get the next term of the progression. The term a is the first term of the progression, n is the number of terms in the progression and d is the common difference. The nth term is normally represented by T_n and the sum to n terms of an A.P. is denoted by S_n

n^{th} term = $T_n = a + (n-1)d$

Sum to n terms = $S_n = \left(\dfrac{n}{2}\right)[2a + (n-1)d]$

The sum to n terms of an A.P. can also be written in a different manner. That is,

sum of n terms = $\left(\dfrac{n}{2}\right)[2a + (n-1)d] = \left(\dfrac{n}{2}\right)[a + \{a + (n-1)d\}]$

But, when there are n terms in an A.P., a is the first term and $\{a + (n-1)d\}$ is the last term. Hence,

$S_n = \left(\dfrac{n}{2}\right)[\text{first term} + \text{last term}]$

The average of all the terms in an A.P. is called the arithmetic mean (A.M.) of the A.P. Since the average of a certain numbers is equal to the {sum of all the number/number of numbers}.

Quant Q Volume 2

A.M. of n terms in A.P. $= \dfrac{S_n}{n} = \dfrac{1}{n}\left(\dfrac{n}{2}\right)$

(First Term + Last Term) $= \dfrac{\text{(First Term + Last Term)}}{2}$

i.e. The A.M. of an A.P. is the average of the first and the last terms of the A.P.

The A.M. of an A.P. can also be obtained by considering any two terms which are EQUIDISTANT from the two ends of the A.P. and taking their average, i.e.

(a) the average of the second term from the beginning and the second term from the end is equal to the A.M. of the A.P.

(b) the average of the third term from the beginning and the third term from the end is also equal to the A.M. of the A.P. and so on.

In general, the average of the kth term from the beginning and the kth term from the end is equal to the A.M. of the A.P.

If the A.M. of an A.P. is known, the sum to n terms of the series (S_n) can be expressed as $S_n = n\,(\text{A.M.})$.

In particular, if three numbers are in arithmetic progression, then the middle number is the A.M. i.e. if a, b and c are in A.P., then b is the A.M. of the three terms and

$b = \dfrac{a+c}{2}$.

If a and b are any two numbers, then their A.M. $= \dfrac{a+b}{2}$.

Note:

(i) If three numbers are in A.P., we can take the three terms to be $(a - d)$, a and $(a + d)$.

(ii) If four numbers are in A.P., we can take the four terms to be $(a - 3d)$, $(a - d)$, $(a + d)$ and $(a + 3d)$. The common difference in this case is 2d and not d.

(iii) If five numbers are in A.P., we can take the five terms to be $(a - 2d)$, $(a - d)$, a, $(a + d)$ and $(a + 2d)$.

Inserting arithmetic mean between two numbers:

When n arithmetic means $a_1, a_2, \ldots, a_n$ are inserted between a and b, then a, $a_1, a_2, \ldots, a_n$, b are in A.P.
$t_1 = a$ and $t_{n+2} = b$ of A.P.

The common difference of the A.P. can be obtained as follows:
Given that, n arithmetic means are there between a and b.
$\therefore a = t_1$ and $b = t_{n+2}$
Let d be the common difference.
$b = t_1 + (n + 1)d$
$b = a + (n + 1)d$
$d = \dfrac{(b - a)}{(n + 1)}$

Some important results:

The sum to n terms of the following series are quite useful and, hence, should be remembered by students.

(i) Sum of the first n natural numbers $= \sum\limits_{1}^{n} i = \dfrac{n(n+1)}{2}$

(ii) Sum of squares of the first n natural numbers $= \sum\limits_{1}^{n} i^2 = \dfrac{n(n+1)(2n+1)}{6}$

(iii) Sum of cubes of first n natural numbers

Example 1:

Find the 14th term of an A.P. whose first term is 3 and the common difference is 2.

Solution:

The nth term of an A.P. is given by $t_n = a + (n - 1)d$, where a is the first term and d is the common difference.
∴ $t_{14} = 3 + (14 - 1)2 = 29$

Example 2:

Find the first term and the common difference of an A.P. if the 3rd term is 6 and the 17th term is 34.

Solution:

If a is the first term and the common difference d, then we have
$a + 2d = 6$ ------- (1)
$a + 16d = 34$ ----- (2)
On subtracting equation (1) from equation (2), we get
$14d = 28$ $d = 2$
Substituting the value of d in equation (1), we get $a = 2$
∴ $a = 2$ and $d = 2$

Example 3:

Find the sum of the first 22 terms of an A.P. whose first term is 4 and the common difference is 4/3.

Solution:

Given that, $a = 4$ and $d = \dfrac{4}{3}$.

We have $S_n = \dfrac{n}{2}[2a + (n - 1)d]$

$S_{22} = \left(\dfrac{22}{2}\right)\left[(2)(4) + (22 - 1)\left(\dfrac{4}{3}\right)\right] = (11)(8 + 28) = 396$

Example 4:

Divide 124 into four parts in such a away that they are in A.P. and the product of the first and the 4th part is 128 less than the product of the 2nd and the 3rd parts.

Solution:

Let the four parts be $(a - 3d)$, $(a - d)$, $(a + d)$ and $(a + 3d)$. The sum of these four parts is 124,
i.e. $4a = 124 \quad a = 31$
$(a - 3d)(a + 3d) = (a - d)(a + d) - 128$
$\quad a^2 - 9d^2 = a^2 - d^2 - 128$
$\quad 8d^2 = 128 \quad d = \pm 4$
As $a = 31$, taking $d = 4$, the four parts are 19, 27, 35 and 43.

Note:

If d is taken as –4, then the same four numbers are obtained, but in decreasing order.

Example 5:

Find the three terms in A.P., whose sum is 36 and product is 960.

Solution:

Let the three terms of an A.P. be $(a - d)$, a and $(a + d)$.
Sum of these terms is $3a$.
$3a = 36 \quad a = 12$
Product of these three terms is
$(a + d) \, a \, (a - d) = 960 \quad (12 + d)(12 - d) = 80$
$\quad 144 - d^2 = 80 \quad d = \pm 8$
Taking $d = 8$, we get the terms as 4, 12 and 20.

Note:

If d is taken as –8, then the same numbers are obtained, but in decreasing order.

Geometric Progression (G.P.):

Numbers are said to be in geometric progression when the ratio of any quantity to the number that follows it is the same. In other words, any term of a G.P. (except the first one) can be obtained by multiplying the previous term by a constant factor.

The constant factor is called the common ratio and is normally represented by r. The first term of a G.P. is generally denoted by a.

A geometric progression can be represented by $a, ar, ar^2, \ldots$ where a is the first term and r is the common ratio of the G.P. n^{th} term of the G.P. is ar^{n-1} i.e. $t_n = ar^{n-1}$

$$\text{Sum to n terms} = S_n = \frac{a(1-r^n)}{1-r} = \frac{a(r^n-1)}{r-1} = \frac{r(ar^{n-1}) - a}{r-1}$$

The sum to n terms of a geometric progression can also be written as

$$S_n = \frac{r(\text{Last term}) - \text{First term}}{r-1}$$

Note:

If n terms $a_1, a_2, a_3, \ldots a_n$ are in G.P., then the geometric mean (G.M.) of these n terms is given by $= \sqrt[n]{a_1 a_2 a_3 \ldots a_n}$

If three terms are in geometric progression, then the middle term is the geometric mean of the G.P., i.e. if a, b and c are in G.P., then b is the geometric mean of the three terms.

If there are two terms a and b, their geometric mean is given by G.M. = $\sqrt{ab}$. We see that the 3 terms a, $\sqrt{ab}$, b are in G.P.

For any two positive numbers a and b, their arithmetic mean is always greater than or equal to their geometric mean, i.e. for any two positive numbers a and b, $\frac{a+b}{2} \geq \sqrt{ab}$.

The equality holds if and only if a = b.

Infinite geometric progression:

If $-1 < r < 1$ (or $|r| < 1$), then the sum of a geometric progression does not increase infinitely but "converges" to a particular value, no matter how many terms of the G.P. we take. The sum of an infinite geometric progression is represented by S_Π and is given by the formula,

$$S_\Pi = \frac{a}{1-r^2}, \text{ if } |r| < 1.$$

Example 6:

Find the 7th term of the G.P. whose first term is 6 and common ratio is 2/3.

Solution:

Given that, $t_1 = 6$ and $r = \frac{2}{3}$

We have $t_n = a \cdot r^{n-1}$

$$t_7 = (6)\left(\frac{2}{3}\right)^6 = \frac{(6)(64)}{729} = \frac{128}{243}$$

Example 7:

Find the common ratio of the G.P. whose first and last terms are 25 and 1/625 respectively and the sum of the G.P. is 19531/625.

Solution:

We know that the sum of a G.P is $\dfrac{\text{first term} - r(\text{last term})}{1-r}$

$$\frac{19531}{625} = \frac{25 - (r/625)}{1-r} \qquad r = 1/5$$

Example 8:

Find three numbers of a G.P. whose sum is 26 and product is 216.

Solution:

Let the three numbers be a/r, a and ar.
Given that,

a/r . a . ar = 216;

$\Rightarrow a^3 = 216; a = 6$

a/r + a + ar = 26

$\Rightarrow 6 + 6r + 6r^2 = 26r$

$\Rightarrow 6r^2 - 20r + 6 = 0$

$\Rightarrow 6r^2 - 18r - 2r + 6 = 0$

$\Rightarrow 6r(r - 3) - 2(r - 3) = 0$

$\Rightarrow r = 1/3$ (or) $r = 3$

Hence the three numbers are 2, 6 and 18 (or) 18, 6 and 2

Example 9:

If $|x| < 1$, then find the sum of the series $2 + 4x + 6x^2 + 8x^3 + \ldots$

Solution:

Let $S = 2 + 4x + 6x^2 + 8x^3 +$ ------- (1)

$xS = 2x + 4x^2 + 6x^3 + \ldots$ (2)

(1) – (2) gives

$S(1 - x) = 2 + 2x + 2x^2 + 2x^3 + \ldots$
$= 2(1 + x + x^2 + \ldots)$

$1 + x + x^2 + \ldots$ is an infinite G.P with $a = 1$, $r = x$ and $|r| = |x| < 1$

$\therefore$ Sum of the series = $1/(1 - x)$

$\therefore S(1 - x) = 2/(1 - x)$

$\therefore S = 2/(1 - x)^2$

Example 10:

Find the sum of the series 1, 2/5, 4/25, 8/125, ... ∏.

Solution:

Given that, $a = 1$, $r = \dfrac{2}{5}$ and $|r| = \left|\dfrac{2}{5}\right| < 1$

$\therefore S_\Pi = \dfrac{a}{1-r} = \dfrac{1}{1-\dfrac{2}{5}} = 5/3$

Note:

When n geometric means are there between a and b, the common ratio of the G.P. can be derived as follows.

Given that, n geometric means are there between a and b.
$\therefore a = t_1$ and $b = t_{n+2}$
Let 'r' be the common ratio
$\quad b = (t_1)(r^{n+1}) \quad b = a\, r^{n+1}$
$\quad r^{n+1} = \dfrac{b}{a}$

$\quad r = \sqrt[(n+1)]{\dfrac{b}{a}}$

Harmonic Progression (H.P.):

A progression is said to be a harmonic progression if the reciprocal of the terms in the progression form an arithmetic progression.
For example, consider the series

$\dfrac{1}{2}, \dfrac{1}{5}, \dfrac{1}{8}, \dfrac{1}{11}, ...$

The progression formed by taking reciprocals of terms of the above series is 2, 5, 8, 11,.... Clearly, these terms form an A.P. whose common difference is 3.

n^{th} term of an H.P:

We know that if a, a + d, a + 2d,....are in A.P., then the nth term of this A.P. is a + (n – 1) d. Its reciprocal is

$$\frac{1}{a+(n-1)d}$$

So, n^{th} term of an H.P. whose first two terms are $\frac{1}{a}$ and $\frac{1}{a+d}$ is $\frac{1}{a+(n-1)d}$

Note:

There is no concise general formula for the sum to n terms of an H.P.

Example 11:

Find the 10th term of the H.P. $\frac{3}{2}, 1, \frac{3}{4}, \frac{3}{5}, \ldots\ldots$

Solution:

The given H.P. is $\frac{3}{2}, 1, \frac{3}{4}, \frac{3}{5}, \ldots\ldots$

The corresponding A.P. is $\frac{2}{3}, 1, \frac{4}{3}, \frac{5}{3}, \ldots\ldots$

Here a = $\frac{2}{3}$; d = $1 - \frac{2}{3} = \frac{1}{3}$

∴ T_{10} of the corresponding A.P. is a + (10 – 1)d = $\frac{2}{3} + (9)\frac{1}{3} = \frac{11}{3}$

Hence required term in H.P. is $\frac{3}{11}$

Harmonic Mean (H.M.):

If three terms are in H.P., then the middle term is the H.M. of other two terms.

The harmonic mean of two terms a and b is given by

H.M. = $\frac{2ab}{a+b}$

Relation between A.M., H.M. and G.M. of two numbers

Let x and y be two numbers

$\therefore$ A.M. = $\frac{x+y}{2}$, G.M. = $\sqrt{xy}$ and H.M. = $\frac{2xy}{x+y}$

(A.M.) (H.M.) = (G.M)^2

Inserting n harmonic means between two numbers:

To insert n H.M.'s between two numbers, we first take the corresponding arithmetic series and insert n arithmetic means, and next, we find the corresponding harmonic series.

This is illustrated by the example below:

Example 12:

Insert three harmonic means between $\frac{1}{12}$ and $\frac{1}{20}$

Solution:

After inserting the harmonic means let the harmonic progression be

$\frac{1}{a}, \frac{1}{a+d}, \frac{1}{a+2d}, \frac{1}{a+3d}, \frac{1}{a+4d}$

As $\frac{1}{a} = \frac{1}{12}$ and $\frac{1}{a+4d} = \frac{1}{20}$ a = 12 and d = 2

$\therefore$ The required harmonic means are $\frac{1}{14}, \frac{1}{16}$ and $\frac{1}{18}$

FORMULA SHEET

1. Area of a triangle

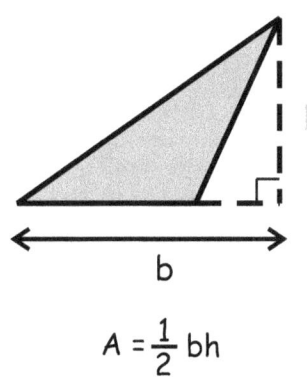

$A = \frac{1}{2} bh$

2. Area of a parellelogram

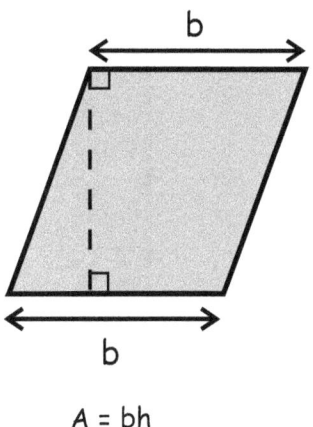

$A = bh$

3. Volume and Surface area of a Cuboid

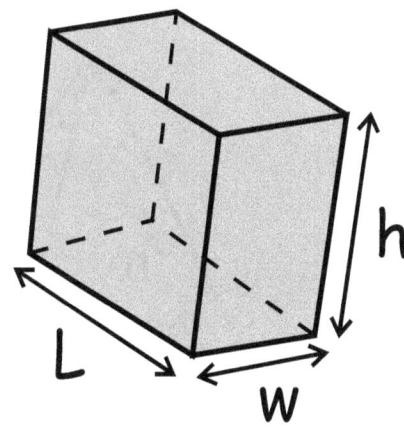

$V = lwh$
$S.A = 2(lw + lh + wh)$

4. volume and Surface area of a Cone

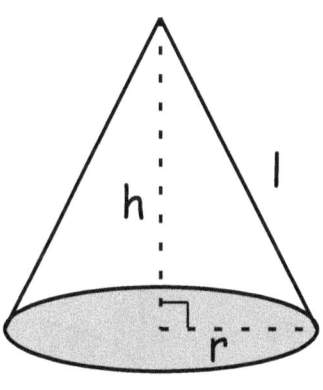

$V = \frac{1}{3} \Pi r^2$

$S.A = \Pi r(l + h)$

5. Perimeter and Area of a Square

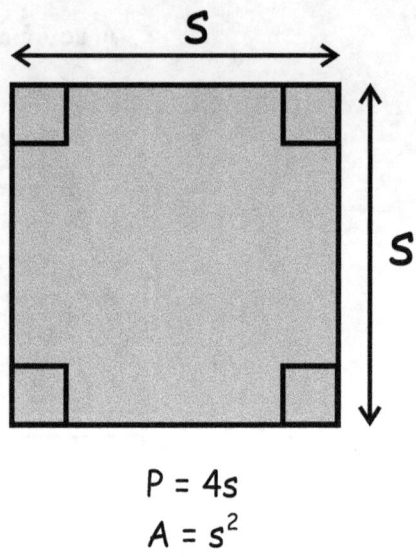

$P = 4s$
$A = s^2$

6. Area of a Trapezium

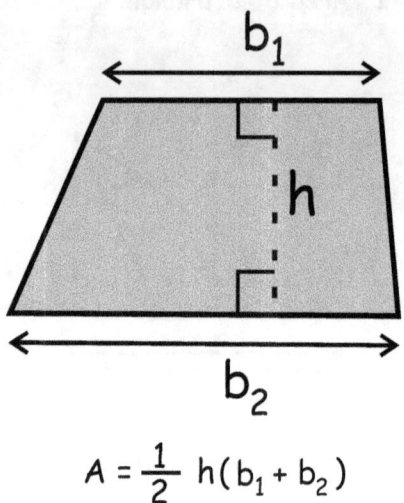

$A = \frac{1}{2} h(b_1 + b_2)$

7. Volume and Surface area of a Cylinder

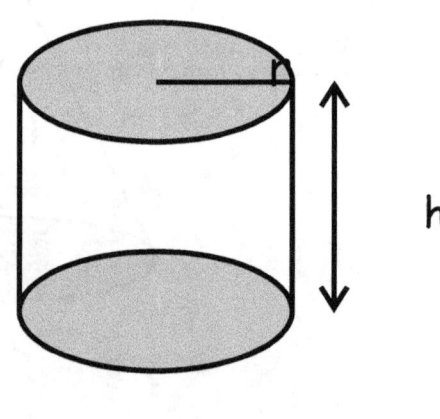

$V = \Pi r^2 h$
$S.A = 2\Pi r(h+r)$

8. Volume and Surface area of a Pyramid

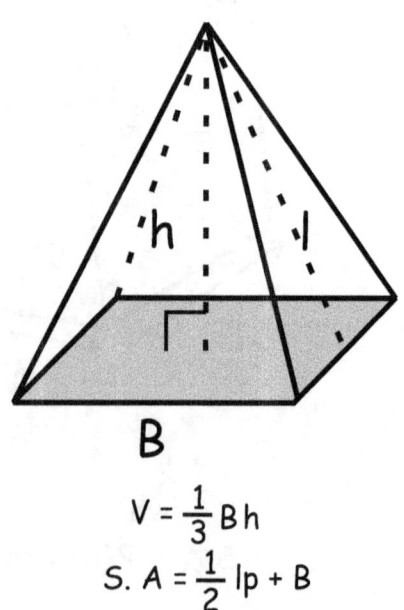

$V = \frac{1}{3} Bh$
$S.A = \frac{1}{2} lp + B$

QUANT - Q

FORMULAE

9. Circumference and Area of a Circle

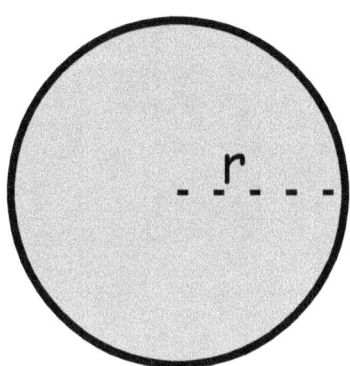

$c = 2\Pi r$
$A = \Pi r^2$

pi
$\Pi = 3.14$
$\Pi = \dfrac{22}{7}$

10. Right angled Triangle (Pythagoran)

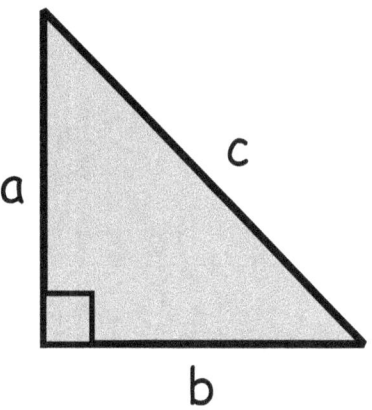

$c^2 = a^2 + b^2$

Pythagorean triplets
Examples : (3 , 4 , 5)
(5 , 12 , 13)
(7 , 24 , 25)
(15 , 20 , 25)
(6 , 8 , 10)
(9 , 12 , 15)
(6 , 8 , 10)
(12 , 16 , 20)
(10 , 24 , 26)

11. Perimeter and Area of a Rectangle

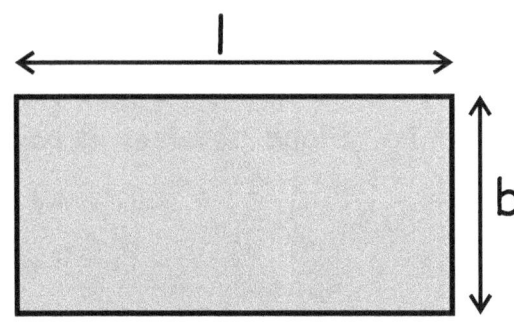

Area = l b ×
Perimeter = 2(l + b)

12. Quadratic formula

$$x = \frac{-b \pm \sqrt{b^2 - 4ac}}{2a}$$

13. Algebraic Identities

$$(a+b)^2 = a^2 + 2ab + b^2$$

$$(a-b)^2 = a^2 - 2ab + b^2$$

$$a^2 - b^2 = (a+b)(a-b)$$

$$a^2 + b^2 = (a+b)^2 - 2ab$$

$$a^2 + b^2 = (a-b)^2 + 2ab$$

14. Equation of a Straight line

Slope intercept form
$y = mx + c$
Where m = slope
c = y - intercept

Point slope form
$y - y_1 = m(x - x_1)$
Where m = slope
Straight line passes through the point (x_1, y_1)

15. Distance between 2 points

$P(x_1, y_1)$ $Q(x_2, y_2)$

$$PQ = \sqrt{(x_2 - x_1)^2 + (y_2 - y_1)^2}$$

16. Slope between 2 points (st line)

$P(x_1, y_1)$ $Q(x_2, y_2)$

$$\text{Slope} = \frac{y_2 - y_1}{x_2 - x_1}$$

Abbreviations

milligram	mg
gram	g
kilogram	kg
milliliter	mL
liter	L
kiloliter	kL
millimeter	mm
centimeter	cm
meter	m
kilometer	km
square centimeter	cm^2
cubic centimeter	cm^3

volume	V
total Square Area	S.A
area of base	B
ounce	oz
pound	lb
quart	qt
gallon	gal.
inches	in.
foot	ft
yard	yd
mile	mi.
square inch	sq in.
square foot	sq ft
cubic inch	cu in.
cubic foot	cu ft

year	yr
month	mon
hour	hr
minute	min
second	sec

QUANT Q
PATTERN RECOGNITION
Vol 2

Prime Factors #1

**QUANT Q
PATTERN RECOGNITION
Vol 2**

Prime Factors #1

Write the prime factorization of each number given below.

1. 55

    (A) 5 · 11   (B) 2 · 2 · 13   (C) 3 · 17   (D) 2 · 3 · 11

2. -31

    (A) -2 · 31   (B) -2 · 2 · 3 · 7 · 31
    (C) -31   (D) -2 · 2 · 3 · 31

3. 12

    (A) 2 · 2   (B) 2 · 2 · 2 · 2
    (C) 2 · 2 · 2 · 2 · 2 · 2   (D) 2 · 2 · 3

4. -75

    (A) -2 · 3 · 3 · 5 · 5   (B) -3 · 5 · 5
    (C) -2 · 2 · 2 · 2 · 3 · 3 · 5 · 5   (D) -2 · 2 · 2 · 2 · 3 · 3 · 3 · 5 · 5

5. -74

    (A) -2 · 3 · 37   (B) -2 · 2 · 2 · 2 · 3 · 37
    (C) -2 · 2 · 2 · 3 · 37   (D) -2 · 37

6. -69

    (A) -3 · 3 · 3 · 3 · 23   (B) -3 · 3 · 3 · 3 · 3 · 23
    (C) -3 · 3 · 23   (D) -3 · 23

7. 40

    (A) 2 · 2 · 2 · 2 · 3   (B) 2 · 2 · 2 · 5
    (C) 3 · 11   (D) 2 · 2 · 3 · 3

Evaluate the below numerical expressions.

1. $(29 - 9) \div ((-1) - 1)$

   A) -27 B) -10
   C) -21 D) -26

2. $14 + 14 - (-4) - 10$

   A) 27 B) 22
   C) 29 D) 7

3. $(-18) + (-8) - 36 \div (-18)$

   A) -21 B) -13
   C) -24 D) -7

4. $20 - (11)(16 \div (-2))$

   A) 108 B) 106
   C) 125 D) 101

5. $((-44) + 15 - 11) \div 20$

   A) -22 B) 0
   C) -2 D) 7

6. $((-25)(2)) \div 10$

   A) -5 B) 3
   C) -1 D) -22

7. $(-19) + 18 - (-7)(6)$

   A) 33 B) 41
   C) 28 D) 23

Evaluate the below expressions.

1. $-11(1 + 7n) - 5(n + 4)$

   (A) -31 - 71n   (B) -31 - 82n
   (C) -31 - 79n   (D) -165n + 167

2. $-2(10 - 12x) - 14(x + 5)$

   (A) -41x + 16   (B) -20x - 130
   (C) -26x - 130   (D) -90 + 10x

3. $-2(v + 5) - 11(-2v + 12)$

   (A) 20v - 142   (B) 33 - 3v
   (C) 33 - v   (D) 33 - 8v

4. $-14(10 + 9n) - 9(13 + 2n)$

   (A) -26n - 30   (B) -9n + 9
   (C) -7n + 9   (D) -257 - 144n

5. $-14(-13 + 10v) - 4(10v + 12)$

   (A) 134 - 180v   (B) -3v + 112
   (C) -17v + 112   (D) 39v - 99

6. $-3(8b + 9) - (7b + 3)$

   (A) -29b - 50   (B) -30b - 50
   (C) 14 - 5b   (D) -31b - 30

7. $-3(7n + 9) - 4(12 + 6n)$

   (A) -67n - 111   (B) -22 - 38n
   (C) 88n + 111   (D) 77n - 111

Evaluate the below expressions.

1. $p^2 - m$; where $m = 1$, and $p = -2$

   (A) 3   (B) 13
   (C) −3   (D) −6

2. $p + 10q$; where $p = -9$, and $q = -4$

   (A) −42   (B) −49
   (C) −56   (D) −53

3. $j + |k|$; where $j = -4$, and $k = 5$

   (A) 9   (B) −20
   (C) −9   (D) 1

4. $p + p + q$; where $p = -9$, and $q = 6$

   (A) −3   (B) −6
   (C) −14   (D) −12

5. $x - 6 + y$; where $x = -3$, and $y = -10$

   (A) −19   (B) −12
   (C) −29   (D) −26

6. $h + j - 9$; where $h = -9$, and $j = 4$

   (A) −8   (B) −11
   (C) −15   (D) −14

7. $|m| - p$; where $m = -2$, and $p = -6$

   (A) 3   (B) 8
   (C) 12   (D) −2

QUANT Q PATTERN RECOGNITION Vol 2

Exponential Expressions #5

Simplify the below with positive exponents.

1. $\left(\dfrac{x^2 y^3}{x^3}\right)^2$

   (A) $\dfrac{x^{20}}{32y^{10}}$ (B) $\dfrac{y^6}{x^2}$

   (C) $\dfrac{x^6}{y^{10}}$ (D) $\dfrac{16x^{12}}{y^3}$

2. $\left(\dfrac{x^5}{(2x^3 y^4)^3}\right)^4$

   (A) $\dfrac{y}{4x^5}$ (B) $\dfrac{1}{4096x^{16}y^{48}}$

   (C) $\dfrac{1}{y^5}$ (D) $2x^3$

3. $\left(\dfrac{2u^3}{u^2 v^2}\right)^2$

   (A) $\dfrac{u^{18}v^7}{4}$ (B) $\dfrac{4u^2}{v^4}$

   (C) $\dfrac{u^4 v^{17}}{2}$ (D) $4u^{14}v^4$

4. $\dfrac{2yx^3}{(2x^4)^2}$

   (A) $\dfrac{1}{x^2}$ (B) $\dfrac{1}{2x^4 y^3}$

   (C) $\dfrac{y}{2x^5}$ (D) $\dfrac{16}{x^{16}y^8}$

5. $\dfrac{(x^4 y^5)^2}{2x^2 y^4}$

   (A) $\dfrac{1}{x^{22}y^8}$ (B) $\dfrac{16y^{16}}{x^4}$

   (C) $4x^6 y^6$ (D) $\dfrac{x^6 y^6}{2}$

6. $\dfrac{2m^2 n^0}{(mn^4)^2}$

   (A) $\dfrac{m^2 n^4}{2}$ B) $m^7 n^9$

   (C) $\dfrac{n^3}{8m^{11}}$ D) $\dfrac{2}{n^8}$

7. $\dfrac{(2x^4 y^2)^3}{2x^2 y^4}$

   (A) $\dfrac{1}{2x^3 y^2}$ (B) $\dfrac{1}{x^3 y^{12}}$

   (C) $4x^{10}y^2$ (D) $x^4 y$

Find the distance between the below points.

1. O(−7, 5), P(−4, −4)

   (A) $2\sqrt{3}$ (B) $2\sqrt{30}$ (C) $12\sqrt{2}$ (D) $3\sqrt{10}$

2. Q(0, 5), R(−2, −1)

   (A) $\sqrt{6}$ (B) $2\sqrt{5}$ (C) $2\sqrt{2}$ (D) $2\sqrt{10}$

3. S(5, −7), T(1, 3)

   (A) $9\sqrt{2}$ (B) $\sqrt{14}$ (C) $2\sqrt{29}$ (D) $2\sqrt{5}$

4. U(4, 2), V(2, −3)

   (A) $\sqrt{37}$ (B) $\sqrt{7}$ (C) $\sqrt{29}$ (D) $\sqrt{35}$

5. W(4, 4), X(−6, 0)

   (A) $\sqrt{6}$ (B) $2\sqrt{29}$ (C) $2\sqrt{5}$ (D) $\sqrt{14}$

6. Y(2, 7), Z(−6, 5)

   (A) $4\sqrt{10}$ (B) $\sqrt{10}$ (C) 4 (D) $2\sqrt{17}$

7. A(7, 7), B(3, 3)

   (A) $\sqrt{8}$ (B) $4\sqrt{2}$ (C) $\sqrt{2}$ (D) $\sqrt{6}$

8. C(−2, −2), D(6, 8)

   (A) $\sqrt{2}$ (B) $3\sqrt{2}$ (C) $2\sqrt{41}$ (D) $2\sqrt{13}$

Find the midpoint of each line segment given in below graphs.

1.

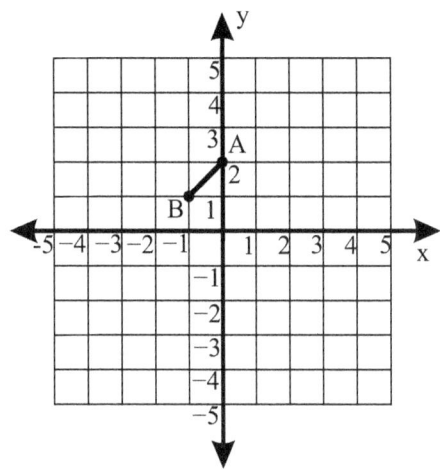

(A) (0.5, 0.5)   (B) (1, 0)

(C) (−2, 0)   (D) (−0.5, 1.5)

2.

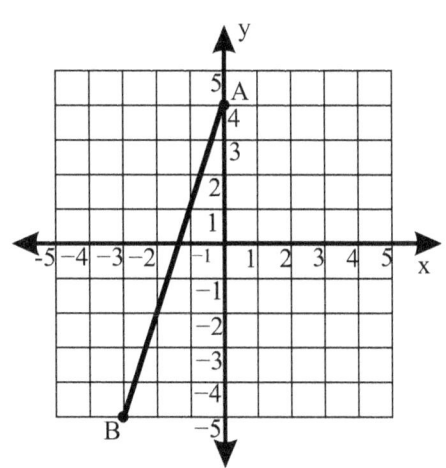

(A) (−1.5, −0.5)   (B) (2, −4)

(C) (1.5, 4.5)   (D) (−6, −14)

Find the midpoint of each line segment given in below graphs.

3.

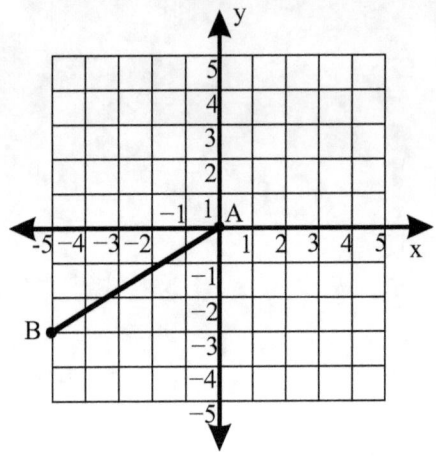

(A) (2.5, 1.5) (B) (0, −4)

(C) (−2.5, −1.5) (D) (−10, −6)

4.

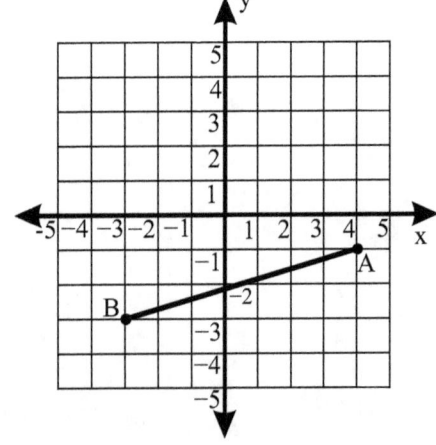

(A) (−10, −5) (B) (3.5, 1)

(C) (1.5, −3) (D) (0.5, −2)

Find the midpoint of each line segment given in below graphs.

5.

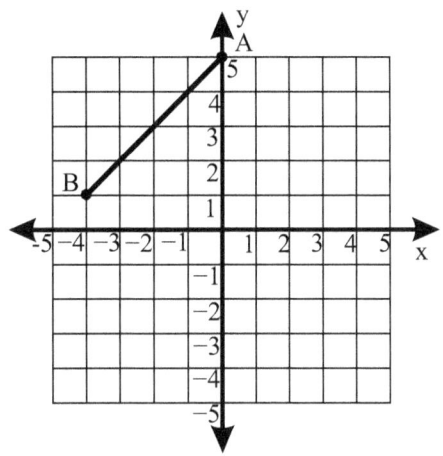

(A)   (−2, 3)   (B)   (2.5, −1.5)
(C)   (2, 2)   (D)   (−8, −3)

6.

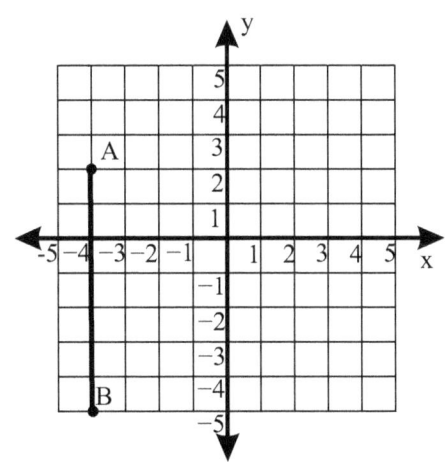

(A)   (−4.5, −1)   (B)   (−4, −1.5)
(C)   (0, −3.5)   (D)   (−4, 9)

Find the midpoint of each line segment given in below graphs.

7.

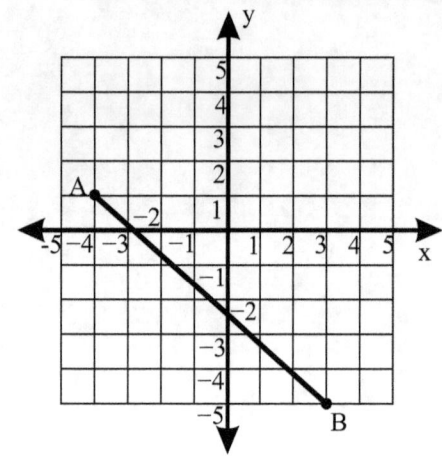

(A) (−0.5, −2) (B) (10, −11)

(C) (−1.5, −1) (D) (−3.5, 3)

Write the standard form of the straight line passing through the given points below.

1. $O(-1, 1)$ and $P(-2, -5)$

   (A) $6x + y = -4$ (B) $6x + y = 7$

   (C) $6x - y = -7$ (D) $3x + y = 4$

2. $Q(-1, 1)$ and $R(2, 3)$

   (A) $x - 3y = 15$ (B) $x + 3y = 15$

   (C) $2x - 3y = -5$ (D) $32x + 15y = -1$

3. $S(1, -4)$ and $T(0, 3)$

   (A) $3x - y = 7$ (B) $7x + y = 3$

   (C) $2x - y = -7$ (D) $2x + y = 2$

4. $U(-4, 0)$ and $V(0, 1)$

   (A) $3x + y = -16$ (B) $4x + y = 16$

   (C) $5x - y = 16$ (D) $x - 4y = -4$

5. $W(2, -5)$ and $X(-2, 1)$

   (A) $3x + 2y = -4$ (B) $x + 3y = 0$

   (C) $3x + 4y = -2$ (D) $2x + 3y = 10$

6. $A(0, 4)$ and $B(5, 1)$

   (A) $3x + 5y = -25$ (B) $3x + 5y = 20$

   (C) $x - 5y = 25$ (D) $3x + 5y = 25$

7. $C(-5, 4)$ and $D(2, 3)$

   (A) $x + 7y = -28$ (B) $x + 7y = -23$

   (C) $x + 7y = 23$ (D) $x + 7y = 14$

8. $E(0, 1)$ and $F(0, -4)$

   (A) $3x - 4y = 0$ (B) $4x = 0$

   (C) $x = 0$ (D) $-4y = 0$

Find the slope of the straight line passing through the points given below.

1. X(−20, 2), Y(−20, 1)

   (A) Undefined (B) −2
   (C) 2 (D) 0

2. C(−13, 16), D(8, 1)

   (A) $\dfrac{7}{5}$ (B) $-\dfrac{7}{5}$
   (C) $\dfrac{5}{7}$ (D) $-\dfrac{5}{7}$

3. E(15, 8), F(17, −19)

   (A) $\dfrac{27}{2}$ (B) $-\dfrac{2}{27}$
   (C) $\dfrac{2}{27}$ (D) $-\dfrac{27}{2}$

Find the slope of the straight line passing through the points given below.

4. $H(-11, 8), I(14, -2)$

   (A) $\dfrac{5}{2}$ (B) $-\dfrac{2}{5}$

   (C) $-\dfrac{5}{2}$ (D) $\dfrac{2}{5}$

5. $J(-7, -14), K(-13, 13)$

   (A) $-\dfrac{2}{9}$ (B) $\dfrac{2}{9}$

   (C) $-\dfrac{9}{2}$ (D) $\dfrac{9}{2}$

6. $L(5, 14), M(-17, -17)$

   (A) $-\dfrac{22}{31}$ (B) $-\dfrac{31}{22}$

   (C) $\dfrac{31}{22}$ (D) $\dfrac{22}{31}$

Find the slope of the straight line passing through the points given below.

7. R(1, 13), S(−9, −4)

(A) $\dfrac{17}{10}$ (B) $\dfrac{10}{17}$

(C) $-\dfrac{10}{17}$ (D) $-\dfrac{17}{10}$

8. N(−6, −15), M(−16, −15)

(A) $-\dfrac{3}{4}$ (B) 0

(C) $\dfrac{3}{4}$ (D) Undefined

9. I(15, −15), J(−5, −7)

(A) $-\dfrac{2}{5}$ (B) $-\dfrac{5}{2}$

(C) $\dfrac{2}{5}$ (D) $\dfrac{5}{2}$

Find the slope-intercept form of the straight line with the given slope and y-intercept for all the problems given below.

1. Slope $= -\dfrac{6}{5}$, y-intercept $= -4$

   (A) $y = -4x - \dfrac{2}{5}$ (B) $y = -\dfrac{2}{5}x - 4$

   (C) $y = \dfrac{4}{5}x - 4$ (D) $y = -\dfrac{6}{5}x - 4$

2. Slope $= -\dfrac{7}{4}$, y-intercept $= -3$

   (A) $y = \dfrac{7}{4}x - 3$ (B) $y = -\dfrac{7}{4}x - 3$

   (C) $y = 3x + \dfrac{7}{4}$ (D) $y = -3x + \dfrac{7}{4}$

3. Slope $= -\dfrac{7}{5}$, y-intercept $= 4$

   (A) $y = -\dfrac{7}{5}x + 4$ (B) $y = 4x + \dfrac{7}{5}$

   (C) $y = -4x + \dfrac{7}{5}$ (D) $y = \dfrac{7}{5}x + 4$

Slope intercept form #10

Find the slope-intercept form of the straight line with the given slope and y-intercept for all the problems given below.

4. Slope = $-\dfrac{2}{3}$, y-intercept = 2

  (A) $y = -\dfrac{2}{3}x + 2$ (B) $y = 2x - \dfrac{1}{3}$

  (C) $y = \dfrac{1}{3}x + 2$ (D) $y = -\dfrac{1}{3}x + 2$

5. Slope = $\dfrac{9}{2}$, y-intercept = -5

  (A) $y = \dfrac{9}{2}x - 5$ (B) $y = -4x - 5$

  (C) $y = \dfrac{1}{2}x - 5$ (D) $y = -2x - 5$

6. Slope = 1, y-intercept = -1

  (A) $y = x - 1$ (B) $y = -x - 1$

  (C) $y = 3x - 1$ (D) $y = -3x - 1$

Find the slope-intercept form of the straight line with the given slope and y-intercept for all the problems given below.

7. Slope = −1, y-intercept = −1

   (A) $y = 3x - 1$ (B) $y = -3x - 1$

   (C) $y = -x - 3$ (D) $y = -x - 1$

8. Slope = $\dfrac{4}{3}$, y-intercept = 1

   (A) $y = \dfrac{4}{3}x + 1$ (B) $y = \dfrac{4}{3}x - 1$

   (C) $y = x + \dfrac{4}{3}$ (D) $y = -x + \dfrac{4}{3}$

Find the slope of the straight line for each of the questions below.

1)
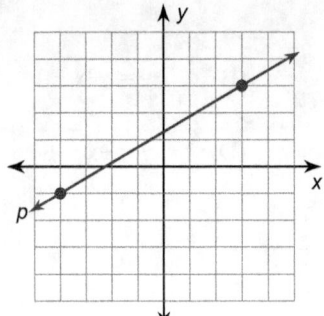

A) $\dfrac{7}{4}$ B) $\dfrac{4}{7}$

C) $-\dfrac{7}{4}$ D) $-\dfrac{4}{7}$

2)
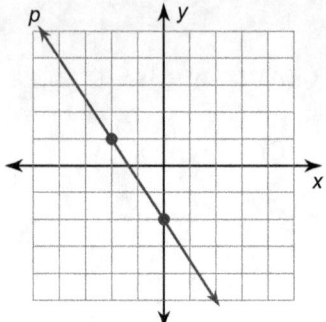

A) $-\dfrac{3}{2}$ B) $\dfrac{3}{2}$

C) $\dfrac{2}{3}$ D) $-\dfrac{2}{3}$

3)
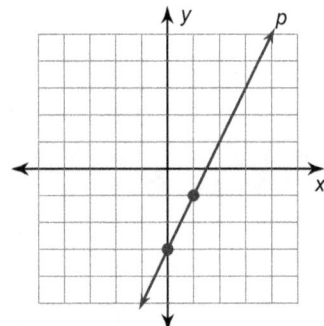

A) 2 B) $-\dfrac{1}{2}$

C) $\dfrac{1}{2}$ D) -2

4)
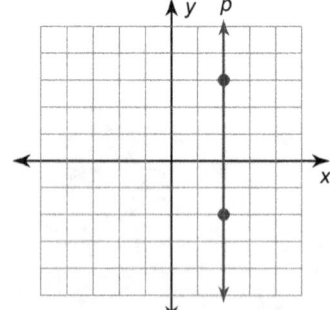

A) $-\dfrac{2}{5}$ B) Undefined

C) $\dfrac{2}{5}$ D) 0

Find the slope of the straight line for each of the questions below.

5)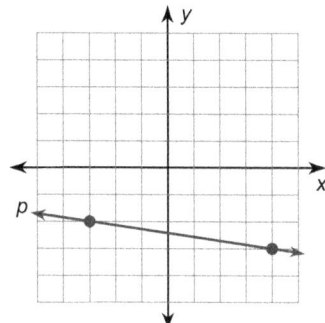

A) $\dfrac{1}{7}$ B) $-\dfrac{1}{7}$

C) 7 D) -7

6)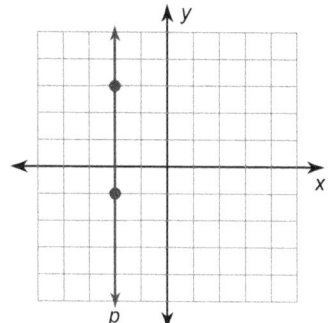

A) 0 B) $-\dfrac{1}{5}$

C) $\dfrac{1}{5}$ D) Undefined

7)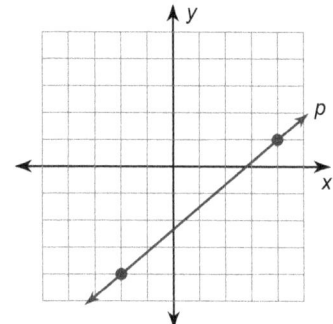

A) $\dfrac{5}{6}$ B) $\dfrac{6}{5}$

C) $-\dfrac{6}{5}$ D) $-\dfrac{5}{6}$

Find the slope of the straight lines given below

1) $7x - 2y = 8$

   A) $-\dfrac{7}{2}$ B) $-\dfrac{2}{7}$

   C) $\dfrac{7}{2}$ D) $\dfrac{2}{7}$

2) $5x - 3y = 0$

   A) $\dfrac{5}{3}$ B) $-\dfrac{5}{3}$

   C) $\dfrac{3}{5}$ D) $-\dfrac{3}{5}$

3) $3x - 2y = -6$

   A) $-\dfrac{2}{3}$ B) $\dfrac{3}{2}$

   C) $\dfrac{2}{3}$ D) $-\dfrac{3}{2}$

4) $8x + 5y = -15$

   A) $\dfrac{8}{5}$ B) $-\dfrac{8}{5}$

   C) $\dfrac{5}{8}$ D) $-\dfrac{5}{8}$

5) $x + 4y = -8$

   A) 4 B) $\dfrac{1}{4}$

   C) -4 D) $-\dfrac{1}{4}$

6) $x + 2y = 0$

   A) 2 B) $-\dfrac{1}{2}$

   C) -2 D) $\dfrac{1}{2}$

7) $5x - 4y = 16$

   A) $-\dfrac{5}{4}$ B) $\dfrac{5}{4}$

   C) $-\dfrac{4}{5}$ D) $\dfrac{4}{5}$

Parallel line slope #13

Find the slope of the straight line parallel to the line given below

1) $x + 5y = 15$

   A) 5   B) -5

   C) $\dfrac{1}{5}$ D) $-\dfrac{1}{5}$

2) $x - y = 0$

   A) 1   B) -1

   C) $\dfrac{5}{3}$ D) $-\dfrac{5}{3}$

3) $7x + 4y = 20$

   A) $-\dfrac{4}{7}$ B) $-\dfrac{7}{4}$

   C) $\dfrac{4}{7}$ D) $\dfrac{7}{4}$

4) $x = 3$

   A) Undefined   B) $\dfrac{3}{5}$

   C) $-\dfrac{3}{5}$ D) 0

5) $x - y = 1$

   A) $\dfrac{2}{5}$ B) $-\dfrac{2}{5}$

   C) 1   D) -1

6) $x + 2y = 6$

   A) $\dfrac{1}{2}$ B) $-\dfrac{1}{2}$

   C) 2   D) -2

7) $x - 4y = 4$

   A) 4   B) $-\dfrac{1}{4}$

   C) $\dfrac{1}{4}$ D) -4

Find the slope of the straight line perpendicular to the lines given below

1) $y = -4x$

A) $\dfrac{1}{4}$ B) 4

C) $-\dfrac{1}{4}$ D) -4

2) $y = -\dfrac{4}{5}x + 1$

A) $\dfrac{5}{4}$ B) $-\dfrac{5}{4}$

C) $-\dfrac{4}{5}$ D) $\dfrac{4}{5}$

3) $y = -\dfrac{1}{3}x + 2$

A) 3 B) -3

C) $-\dfrac{1}{3}$ D) $\dfrac{1}{3}$

4) $y = \dfrac{1}{2}x$

A) 2 B) $\dfrac{1}{2}$

C) $-\dfrac{1}{2}$ D) -2

5) $y = -\dfrac{3}{4}x + 5$

A) $\dfrac{4}{3}$ B) $-\dfrac{3}{4}$

C) $\dfrac{3}{4}$ D) $-\dfrac{4}{3}$

6) $y = x + 3$

A) $-\dfrac{1}{2}$ B) 1

C) $\dfrac{1}{2}$ D) -1

7) $y = \dfrac{5}{2}x - 2$

A) $\dfrac{2}{5}$ B) $-\dfrac{2}{5}$

C) $\dfrac{5}{2}$ D) $-\dfrac{5}{2}$

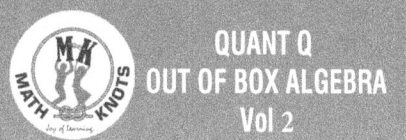

RADICALS 1
#15

Simplify the below radicals.

1. $\dfrac{4}{\sqrt{6}-\sqrt{2}}$

   (A) $\dfrac{2\sqrt{6}}{3}$ (B) $\sqrt{6}+\sqrt{2}$

   (C) $\sqrt{6}+\sqrt{3}$ (D) $\dfrac{4\sqrt{5}+4\sqrt{2}}{39}$

2. $\dfrac{-3+8\sqrt{2}}{8-\sqrt{6}}$

   (A) $\dfrac{-24+3\sqrt{6}+64\sqrt{2}-16\sqrt{3}}{58}$ (B) $3-6\sqrt{5}$

   (C) $\dfrac{-17+58\sqrt{2}+16\sqrt{6}}{63}$ (D) $\dfrac{-24-3\sqrt{6}+64\sqrt{2}+16\sqrt{3}}{58}$

3. $\dfrac{4}{5+5\sqrt{3}}$

   (A) $\dfrac{-3+3\sqrt{3}}{5}$ (B) $-10+6\sqrt{3}$

   (C) $\dfrac{-2+2\sqrt{3}}{5}$ (D) $\dfrac{-1+\sqrt{3}}{5}$

Simplify the below radicals.

4. $\dfrac{9}{7\sqrt{10}-6}$

   (A) $\dfrac{4\sqrt{10}-54}{214}$ (B) $\dfrac{63\sqrt{10}-54}{454}$

   (C) $\dfrac{45\sqrt{10}-81}{169}$ (D) $\dfrac{36\sqrt{10}-27}{302}$

5. $\dfrac{6}{\sqrt{5}-3\sqrt{7}}$

   (A) $\dfrac{-12\sqrt{2}-30\sqrt{7}}{167}$ (B) $\dfrac{-3\sqrt{5}-9\sqrt{7}}{29}$

   (C) $\dfrac{-4\sqrt{5}-12\sqrt{7}}{29}$ (D) $\dfrac{-6\sqrt{5}-18\sqrt{6}}{49}$

6. $\dfrac{9-\sqrt{3}}{7+\sqrt{7}}$

   (A) $\dfrac{18-9\sqrt{2}-2\sqrt{3}+\sqrt{6}}{4}$ (B) $\dfrac{63-9\sqrt{7}+14\sqrt{3}-2\sqrt{21}}{42}$

   (C) $\dfrac{63-9\sqrt{7}-7\sqrt{3}+\sqrt{21}}{42}$ (D) $\dfrac{63-9\sqrt{7}+\sqrt{31}}{34}$

Simplify the below radicals.

7. $\dfrac{5}{8 - 3\sqrt{6}}$

(A) $\dfrac{-20 - 15\sqrt{6}}{76}$ (B) $\dfrac{1 - 3\sqrt{6}}{8}$

(C) $\dfrac{8 - 3\sqrt{6}}{2}$ (D) $\dfrac{21 + 6\sqrt{6}}{25}$

8. $\dfrac{5 - \sqrt{5}}{8 + 3\sqrt{2}}$

(A) $\dfrac{40 - 15\sqrt{2} - 32\sqrt{5} + 12\sqrt{10}}{46}$ (B) $\dfrac{48 - 18\sqrt{2} - 8\sqrt{5} + 3\sqrt{10}}{46}$

(C) $\dfrac{40 - 15\sqrt{2} - 8\sqrt{5} + 3\sqrt{10}}{46}$ (D) $\dfrac{41 - 21\sqrt{2} - 8\sqrt{6}}{46}$

RADICALS 2
#16

Simplify the below to the lowest positive terms.

1. $\dfrac{8n - 3\sqrt{6n^4}}{\sqrt{28n}}$

   (A) $\dfrac{8\sqrt{7n} - 3n\sqrt{42n}}{14}$ (B) $\dfrac{8\sqrt{10n} - 21n\sqrt{n}}{14}$

   (C) $\dfrac{5\sqrt{7n} - 3n\sqrt{42n}}{14}$ (D) $\dfrac{7\sqrt{7n} - 3n\sqrt{42n}}{14}$

2. $\dfrac{9x}{10\sqrt{10x^2}}$

   (A) $\dfrac{9\sqrt{10}}{100}$ (B) $\dfrac{3\sqrt{10}}{50}$

   (C) $\dfrac{3\sqrt{10}}{40}$ (D) $\dfrac{9\sqrt{10}}{70}$

3. $\dfrac{-10r + 4\sqrt{7r}}{12r}$

   (A) $\dfrac{-8\sqrt{3r} + 8}{3}$ (B) $\dfrac{-5\sqrt{3r} + 2\sqrt{21}}{3}$

   (C) $\dfrac{-14\sqrt{3r} - 4\sqrt{26}}{3}$ (D) $\dfrac{-4\sqrt{14r} + 14\sqrt{2}}{7}$

RADICALS 2
#16

Simplify the below to the lowest positive terms.

4. $$\dfrac{8r + \sqrt{r^3}}{\sqrt{35r^3}}$$

(A) $\dfrac{24\sqrt{5r} + 5r}{34r}$ (B) $\dfrac{7\sqrt{35r} + r\sqrt{105}}{35r}$

(C) $\dfrac{8\sqrt{35r} + r\sqrt{35}}{35r}$ (D) $\dfrac{8\sqrt{38r} + r\sqrt{38}}{38r}$

5. $$\dfrac{\sqrt{5x^4y^2} + 6y^2}{\sqrt{29x^3y^3}}$$

(A) $\dfrac{x^2\sqrt{145xy} + 6y\sqrt{29xy}}{29x^2y}$ (B) $\dfrac{x^2\sqrt{174xy} + 6y\sqrt{29xy}}{29x^2y}$

(C) $\dfrac{x^2\sqrt{15xy} + 6y\sqrt{3xy}}{9x^2y}$ (D) $\dfrac{x^2\sqrt{145xy} + 5y\sqrt{29xy}}{29x^2y}$

6. $$\dfrac{10 - \sqrt{6x}}{6\sqrt{39x^2}}$$

(A) $\dfrac{10\sqrt{39} - 3\sqrt{26x}}{234x}$ (B) $\dfrac{5\sqrt{39} - 7\sqrt{6x}}{117x}$

(C) $\dfrac{5\sqrt{39} + 2\sqrt{39x}}{117x}$ (D) $\dfrac{10\sqrt{41} - \sqrt{246x}}{164x}$

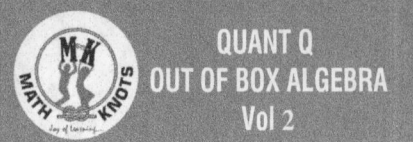

Simplify the below to the lowest positive terms.

7. $\dfrac{6+\sqrt{5n^4}}{8\sqrt{28n^4}}$

(A) $\dfrac{3\sqrt{7}+n^2\sqrt{39}}{112n^2}$ (B) $\dfrac{6\sqrt{7}+n^2\sqrt{35}}{112n^2}$

(C) $\dfrac{3\sqrt{7}+n^2\sqrt{7}}{56n^2}$ (D) $\dfrac{9\sqrt{7}+n^2\sqrt{35}}{112n^2}$

8. $\dfrac{3\sqrt{2n^4}}{10\sqrt{7n^2}}$

(A) $\dfrac{3n\sqrt{7}}{35}$ (B) $\dfrac{3n\sqrt{14}}{77}$

(C) $\dfrac{3n\sqrt{14}}{70}$ (D) $\dfrac{2n\sqrt{14}}{35}$

Solve the below inequalities.

1. $396 \geq -6v + 6(2 - 7v)$

   (A) $v \geq 1$ (B) $v \geq -8$

   (C) $v \leq -26$ (D) $v \leq 1$

2. $3(-7a - 7) \leq -84$

   (A) $a \geq -22$ (B)   All real numbers

   (C) $a \leq 3$ (D) $a \geq 3$

3. $-7 - 3(1 + 4r) \leq -106$

   (A) $r \geq 8$ (B) $r \geq 0$

   (C) $r \geq -13$ (D) $r \geq -10$

4. $2(-1 - 8k) < -82$

   (A) $k < -14$ (B) $k > -1$

   (C) $k > -14$ (D) $k > 5$

5. $4(6 - 8r) \geq 248$

   (A) $r \leq -22$ (B) $r \geq -1$

   (C) $r \leq -7$ (D) $r \geq -22$

6. $-82 < 8(r - 5) - 2r$

   (A) $r < -26$ (B) $r < -7$

   (C) $r > -7$ (D) $r < -30$

1. Amy wants to buy a pair of shoes for $87. She gives the cashier $90. How much change does she get back ?

   (A) $177   (B) $3

   (C) $2   (D) $174

2. Jasmine is cooking a casserole. The recipe calls for $3\frac{3}{5}$ cups of rice. She accidentally put in $4\frac{2}{3}$ cups. How many extra cups did she put in ?

   (A) $8\frac{4}{15}$ (B) $1\frac{1}{15}$

   (C) $4\frac{2}{3}$ (D) $\frac{27}{35}$

3. Cynthia will be 58 years old in ten years. How old is she now ?

   (A) 68   (B) 38

   (C) 48   (D) 44

4. Dan said, he will be 87 years old in thirteen years. How old is he now ?

   (A) 61   (B) 74

   (C) 100   (D) 83

5. Kiran ate 12 cookies and realizes he ate $\frac{2}{5}$ of all he had. How many cookies does he had originally ?

   (A) 30   (B) 4.8

   (C) 26   (D) 24

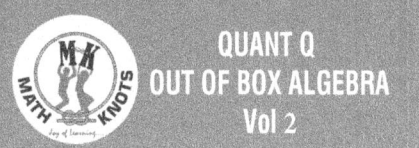

6. How old is Claire if she was 30 years old seventeen years ago ?

(A) 47 (B) 13

(C) 64 (D) 54

7. A recipe for a cake calls for $3\frac{2}{7}$ cups of flour. John accidentally put in $4\frac{1}{9}$ cups. How many extra cups did he put in ?

(A) $\dfrac{207}{259}$ (B) $7\dfrac{25}{63}$

(C) $4\dfrac{1}{9}$ (D) $\dfrac{52}{63}$

8. Susan wants to buy a gift for her friend's birthday. She has $6 which is $\dfrac{2}{3}$ of the gift cost. How much did the gift cost ?

(A) $9 (B) $5

(C) $7 (D) $8

1. Find the area of the circle given diameter = 20 yd.

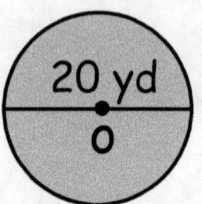

(A) 108Π yd^2 (B) 100Π yd^2

(C) 121Π yd^2 (D) 102Π yd^2

2. Find the area of the circle given diameter = 18 cm.

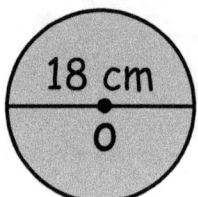

(A) 81Π cm^2 (B) 372Π cm^2

(C) 324Π cm^2 (D) 339Π cm^2

3. Find the area of the circle given diameter = 16 mi.

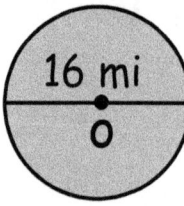

(A) 64Π mi^2 (B) 8Π mi^2

(C) 12Π mi^2 (D) 13Π mi^2

4. Find the circumference of the circle given radius = 7.6 in and round the answer to the nearest tenth.

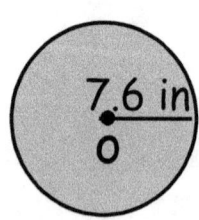

(A) 28.7 in (B) 17.3 in

(C) 23 in (D) 47.7 in

5. Find the circumference of the circle given radius = 3.1 ft and round the answer to the nearest tenth.

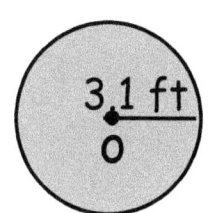

(A) 19.5 ft (B) 18.8 ft

(C) 21.4 ft (D) 19.0 ft

6. Find the circumference of the circle given radius = 10.1 yd and round the answer to the nearest tenth.

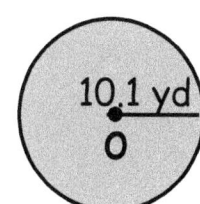

(A) 64.8 yd (B) 72.3 yd

(C) 67.3 yd (D) 63.4 yd

7. Find the circumference of the circle given radius = 8.8 mi and round the answer to the nearest tenth.

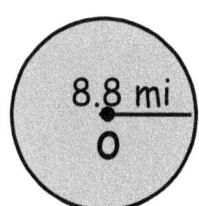

(A) 34 mi (B) 184 mi

(C) 55.3 mi (D) 27.7 mi

1. Find the volume of the sphere with a diameter of 28.6 km and round the answer to the nearest tenth.

   (A)   12248.9 km³   (B)   10901.5 km³

   (C)   21803 km³   (D)   23983.3 km³

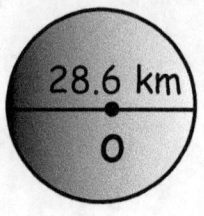

2. Find the volume of the sphere with a diameter of 4.8 cm and round the answer to the nearest tenth.

   (A)   77 cm³   (B)   48.1 cm³

   (C)   57.9 cm³   (D)   96.2 cm³

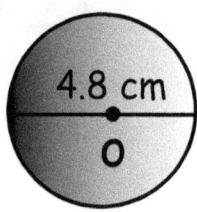

3. Find the volume of the sphere with a radius of 14.7 cm and round the answer to the nearest tenth.

   (A)   6652.9 cm³   (B)   3729.5 cm³

   (C)   7917 cm³   (D)   13305.8 cm³

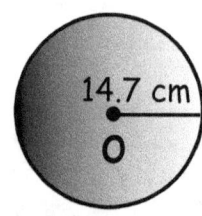

4. Find the volume of the sphere with a radius of 9.3 km and round the answer to the nearest tenth.

   (A)   7816.8 km³   (B)   3369.3 km³

   (C)   9380.2 km³   (D)   3908.4 km³

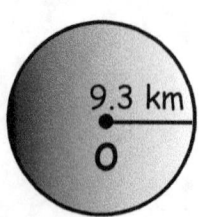

5. Find the volume of the sphere with a radius of 9 ft and round the answer to the nearest tenth.

(A) 8147.1 ft³ (B) 7084.4 ft³

(C) 3053.6 ft³ (D) 3542.2 ft³

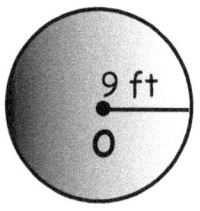

6. Find the volume of the sphere with a radius of 5.6 in and round the answer to the nearest tenth.

(A) 632.6 in³ (B) 456.5 in³

(C) 518.7 in³ (D) 735.6 in³

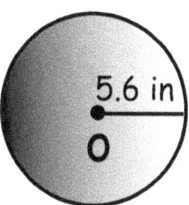

7. Find the volume of the sphere with a diameter of 26 ft and round the answer to the nearest tenth.

(A) 9202.8 ft³ (B) 4737.6 ft³

(C) 5383.6 ft³ (D) 10767.2 ft³

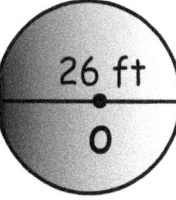

1. Find the volume of a square prism measuring 3 km along each edge of the base and 6 km tall. Round the answer to the nearest tenth.

   (A)   16 km³   (B)   31.9 km³

   (C)   63.7 km³   (D)   54 km³

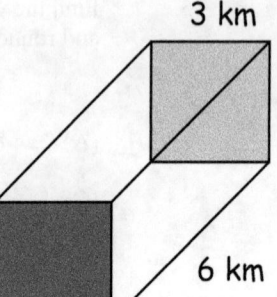

2. Find the volume of a square prism measuring 13 mm along each edge of the base and 6 mm tall.
   Round the answer to the nearest tenth.

   (A)   1216.8 mm³   (B)   1058.6 mm³

   (C)   1014 mm³   (D)   942.2 mm³

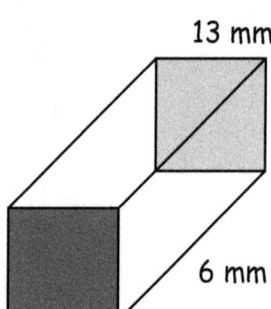

3. Find the volume of a rectangular prism measuring 4 yd and 19 yd along the base and 5 yd tall. Round the answer to the nearest tenth.

   (A)   380 yd³   (B)   334.4 yd³

   (C)   250.2 yd³   (D)   294.3 yd³

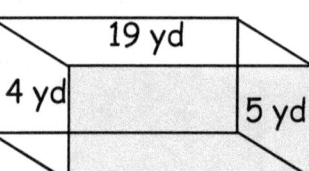

4. Find the volume of a rectangular prism measuring 9 cm and 5 cm along the base and 9cm in tall. Round the answer to the nearest tenth.

(A) 624 cm
(B) 265.7 cm^3
(C) 405 cm
(D) 236.5 cm^3

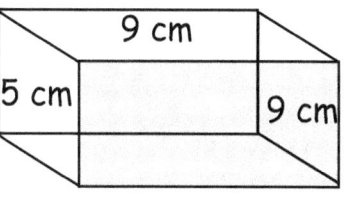

5. Find the volume of a rectangular prism measuring 20 mm and 10 mm along the base and 9 mm tall. Round the answer to the nearest tenth.

(A) 2880 mm^3
(B) 5760 mm^3
(C) 3600 mm^3
(D) 1800 mm^3

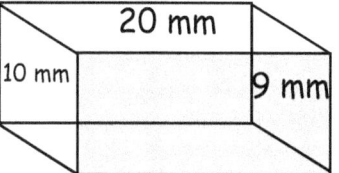

6. Find the volume of a square prism measuring 12 in along each edge of the base and 3 in tall. Round the answer to the nearest tenth.

(A) 488.2 in^3
(B) 466 in^3
(C) 541.9 in^3
(D) 432 in^3

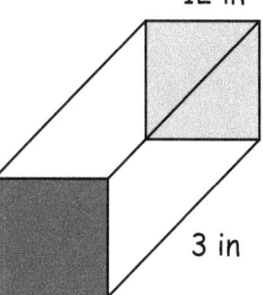

1. Find the volume of a cylinder with a diameter of 10 yd and a height of 3 yd. Round the answer to the nearest tenth.

   (A)   228 yd^3 (B)   235.6 yd^3
   (C)   193.2 yd^3 (D)   193.8 yd^3

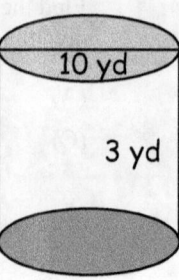

2. Find the volume of a cylinder with a diameter of 40 ft and a height of 11 ft. Round the answer to the nearest tenth.

   (A)   11749.6 ft^3 (B)   9869.7 ft^3
   (C)   13823 ft^3 (D)   4934.9 ft^3

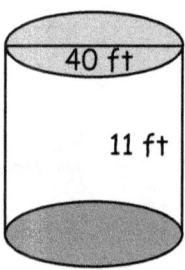

3. Find the volume of a cone with a diameter of 6 km and a height of 11 km. Round the answer to the nearest tenth.

   (A)   103.7 km^3 (B)   51.8 km^3
   (C)   25.9 km^3 (D)   21.8 km^3

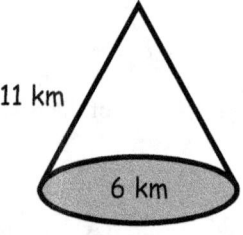

4. Find the volume of a cone with a diameter of 6 m and a height of 8 m. Round the answer to the nearest tenth.

(A) 65.6 m³ (B) 91.4 m³
(C) 75.4 m³ (D) 78.1 m³

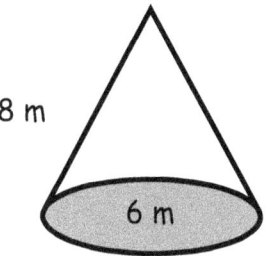

5. Find the volume of a cylinder with a radius of 20 in and a height of 16 in. Round the answer to the nearest tenth.

(A) 20166.5 in³ (B) 20106.2 in³
(C) 23725.3 in³ (D) 24199.8 in³

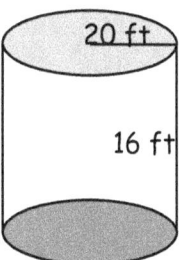

6. Find the volume of a cylinder with a diameter of 16 m and a height of 7 m. Round the answer to the nearest tenth.

(A) 1407.4 m³ (B) 757.2 m³
(C) 934.8 m³ (D) 1140 m³

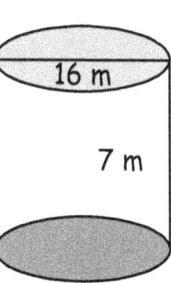

7. Find the volume of a cylinder with a radius of 9 ft and a height of 13 ft. Round the answer to the nearest tenth.

(A) 3308.1 ft³ (B) 3804.3 ft³
(C) 6695.6 ft³ (D) 3347.8 ft³

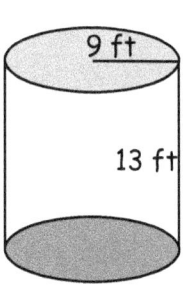

Find the value of 'x' for each of the below:

1)

A) 2 B) 14
C) 5 D) 10

2)

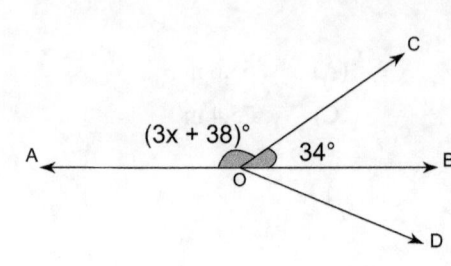

A) 36 B) 40
C) 38 D) 44

3)

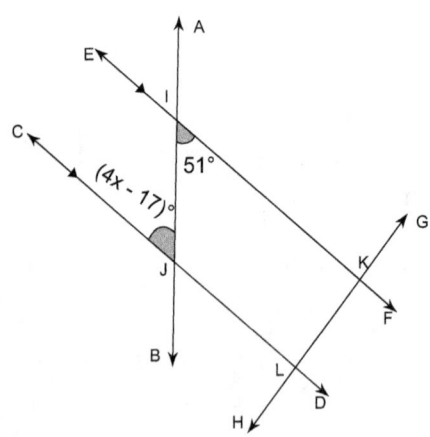

A) 9   B) 17
C) 25 D) 22

4)

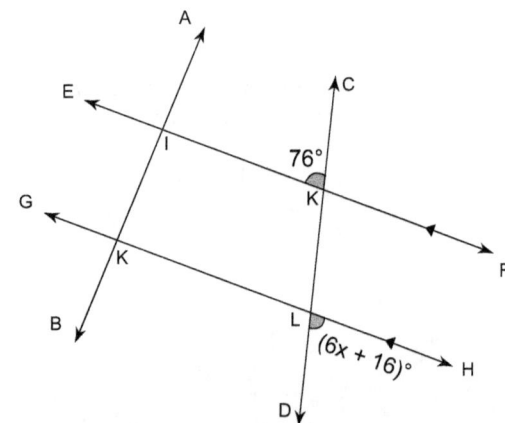

A) 10 B) 6
C) 11 D) -1

5)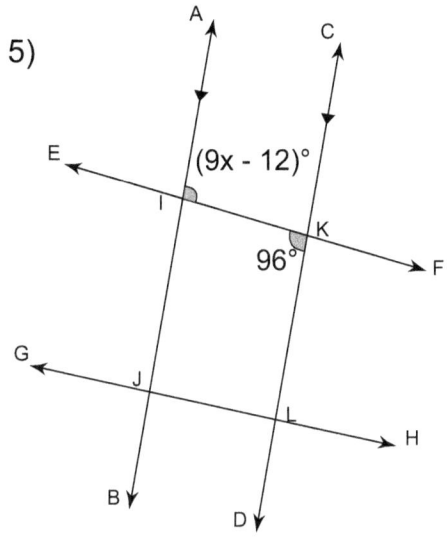

A) 13 B) 12
C) 11 D) 5

6)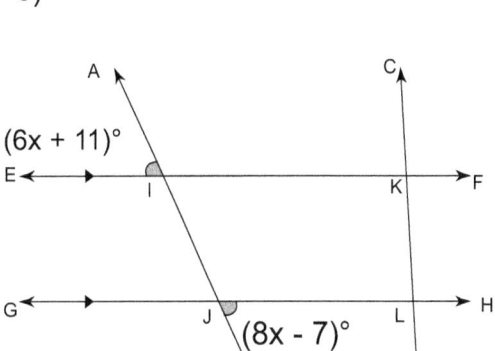

A) 14 B) 9
C) 18 D) 16

7)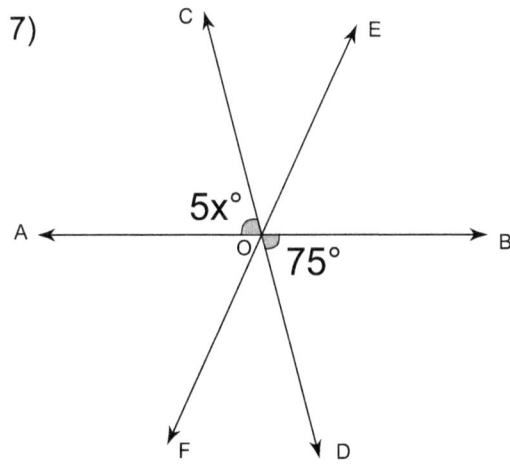

A) 18 B) 12
C) 6   D) 15

Find the value of 'x' for each of the below:

1)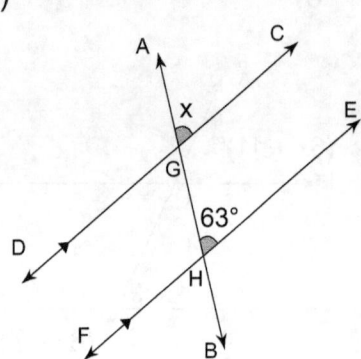

A) 63° B) 87°
C) 117° D) 78°

2)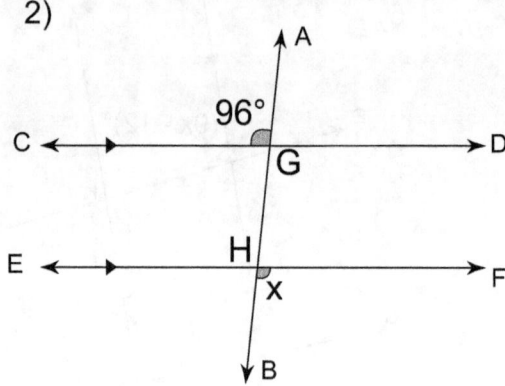

A) 61° B) 84°
C) 96° D) 53°

3)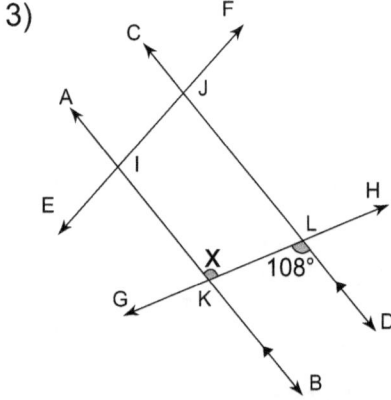

A) 108° B) 18°
C) 162° D) 105°

4)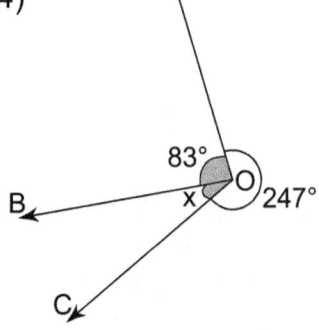

A) 155° B) 30°
C) 150° D) 25°

5)

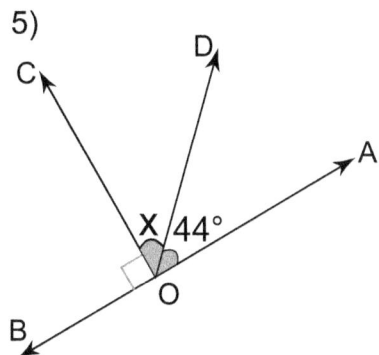

A) 30°   B) 134°

C) 46°   D) 60°

6)

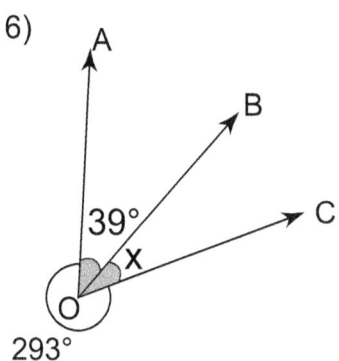

A) 137°   B) 28°

C) 133°   D) 47°

7)

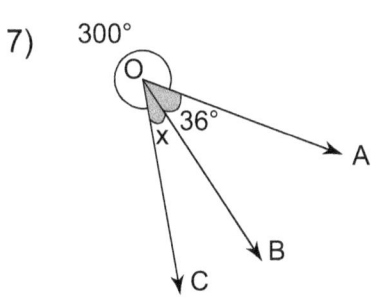

A) 150°   B) 156°

C) 24°   D) 30°

Describe the transformation and find the transformation rule.

1)

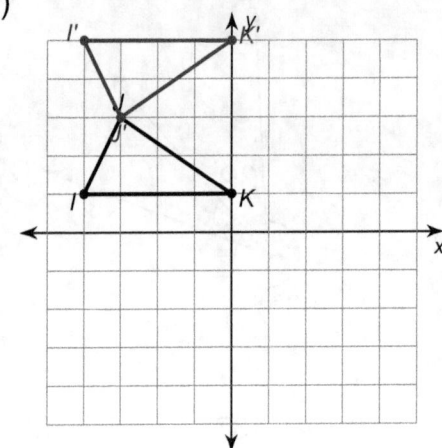

A) Reflection across $x = -3$
B) Reflection across $y = 3$
C) Reflection across $y = -1$
D) Reflection across the y-axis

2)

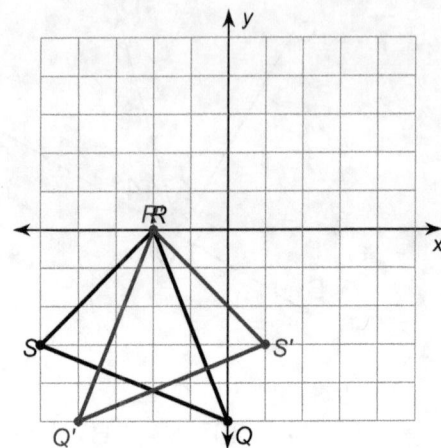

A) Reflection across $y = -2$
B) Reflection across the y-axis
C) Reflection across $x = -2$
D) Reflection across $y = 2$

3)

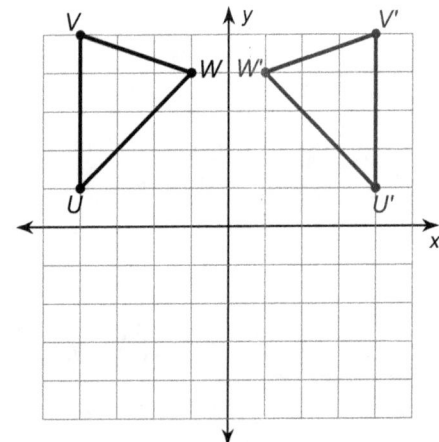

A) Reflection across $x = -1$
B) Reflection across the y-axis ; $y = 4$
C) Reflection across the x-axis
D) Reflection across $y = -1$

4)

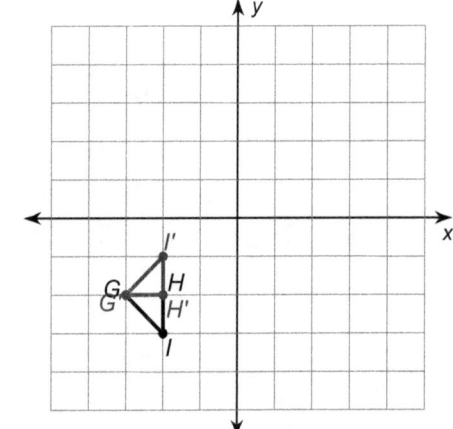

A) Reflection across $y = -2$
B) Reflection across $y = 2$
C) Reflection across $x = -2$
D) Reflection across $x = -1$

Describe the transformation and find the transformation rule.

5)

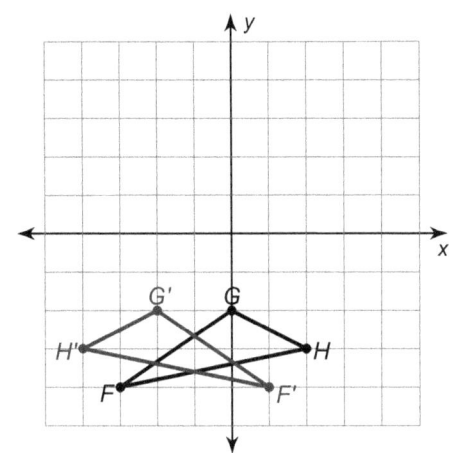

A) Reflection across $y = -2$
B) Reflection across $y = -1$
C) Reflection across $x = -1$
D) Reflection across the y-axis

6)

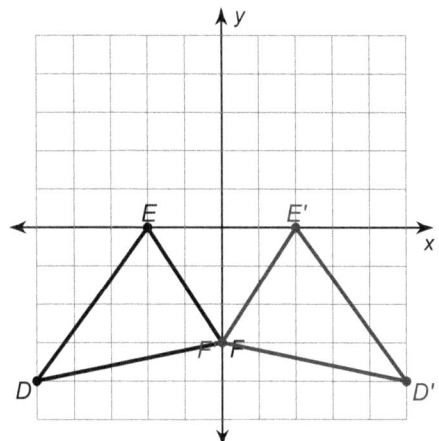

A) Reflection across $y = -4$
B) Reflection across $x = 2$
C) Reflection across $x = 4$
D) Reflection across the y-axis ; $y = -3$

7)

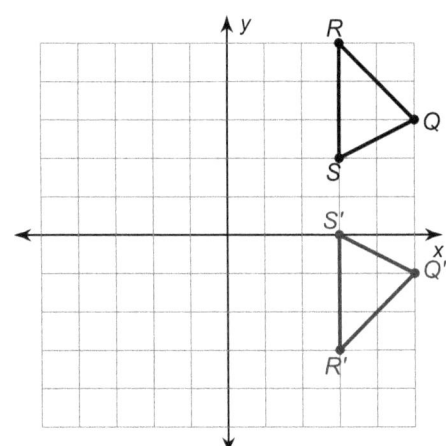

A) Reflection across $y = 1$
B) Reflection across $y = 2$
C) Reflection across $y = 8$
D) Reflection across $x = 4$

Describe the transformation and find the transformation rule.

1)

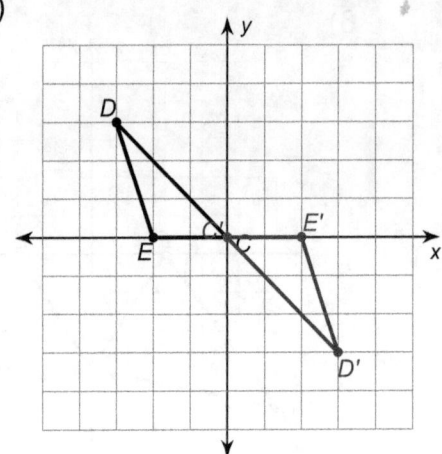

A) Rotation 180° counterclockwise about the origin
B) Rotation 90° clockwise about the origin
C) Translation: 4 units right
D) Rotation 180° about the origin

2)

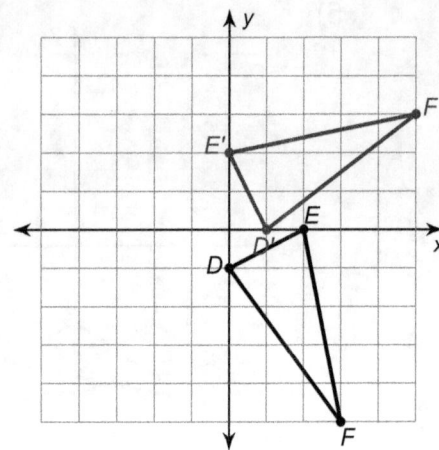

A) Translation: 1 unit right and 3 units up
B) Rotation 270° counterclockwise about the origin
C) Rotation 90° clockwise about the origin
D) Rotation 180° about the origin

3)

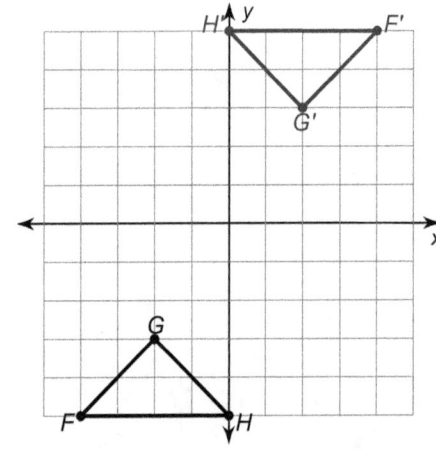

A) Translation: 5 units right and 6 units up
B) Rotation 90° about the origin
C) Rotation 180° counterclockwise about the origin
D) Rotation 90° clockwise about the origin

4)

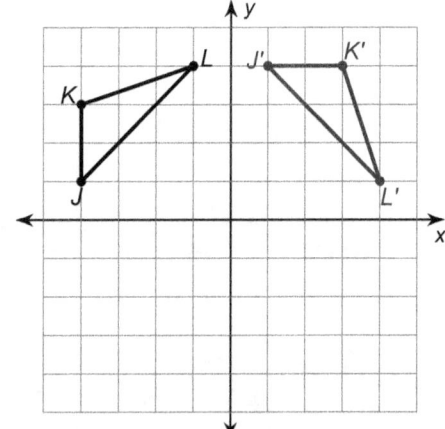

A) Rotation 270° about the origin
B) Rotation 90° clockwise about the origin
C) Translation: 2 units left and 2 units down
D) Rotation 270° clockwise about the origin

Describe the transformation and find the transformation rule.

5)

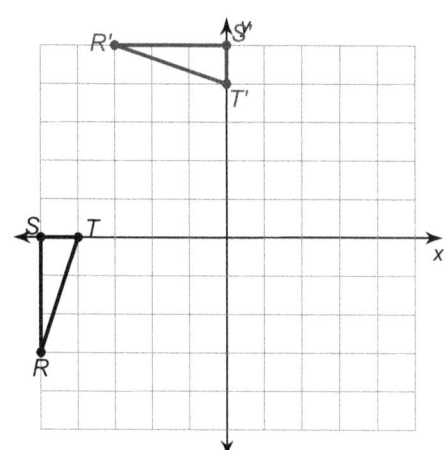

A) Rotation 270° about the origin
B) Reflection across $y = -4$
C) Rotation 180° counterclockwise about the origin
D) Rotation 90° clockwise about the origin

6)

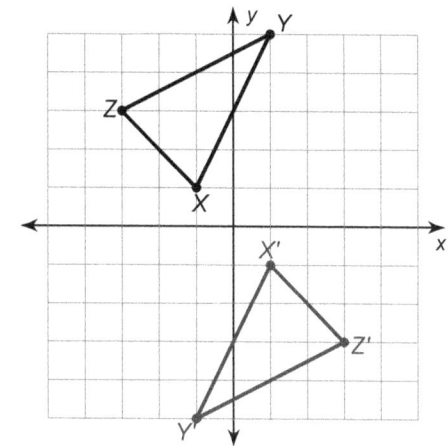

A) Rotation 180° about the origin
B) Reflection across $x = 1$
C) Rotation 90° counterclockwise about the origin
D) Reflection across $x = -1$

7)

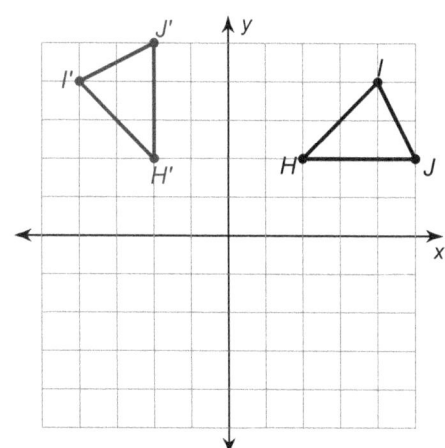

A) Reflection across the y-axis
B) Rotation 90° counterclockwise about the origin
C) Rotation 270° about the origin
D) Rotation 180° clockwise about the origin

1)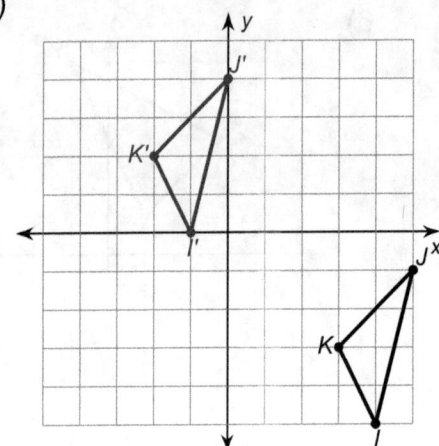

A) Translation: 5 units right and 3 units down
B) Translation: 4 units left and 5 units up
C) Translation: 5 units left and 5 units up
D) Translation: 3 units right and 3 units down

2)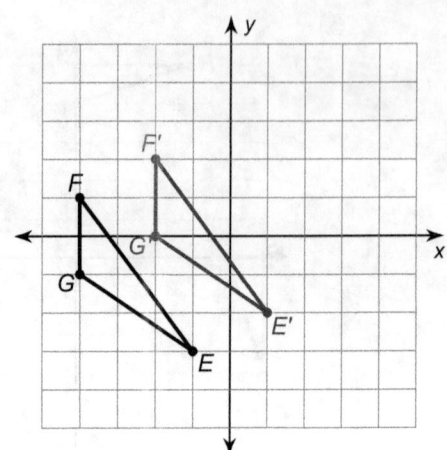

A) Translation: 3 units right
B) Translation: 2 units left
C) Translation: 5 units right and 2 units down
D) Translation: 2 units right and 1 unit up

3)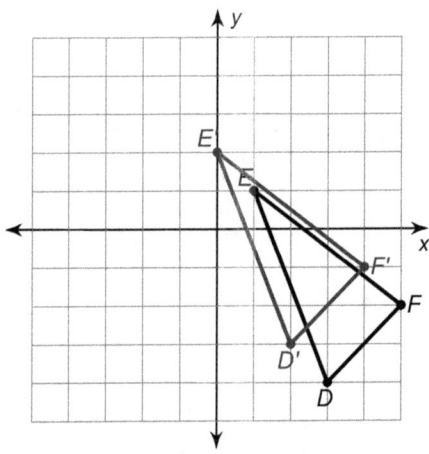

A) Translation: 1 unit right and 2 units down
B) Translation: 1 unit left and 1 unit up
C) Translation: 2 units right
D) Translation: 2 units right and 3 units down

4)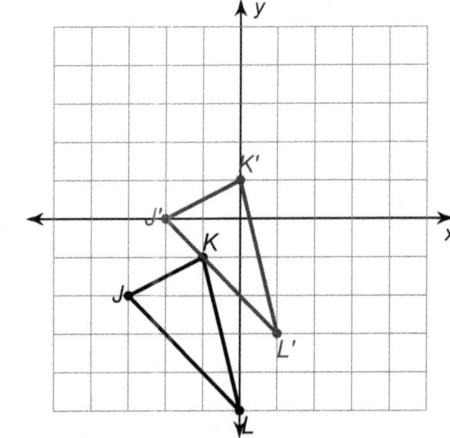

A) Translation: 5 units left
B) Translation: 1 unit right and 2 units up
C) Translation: 3 units right and 2 units down
D) Translation: 3 units left and 2 units up

5)

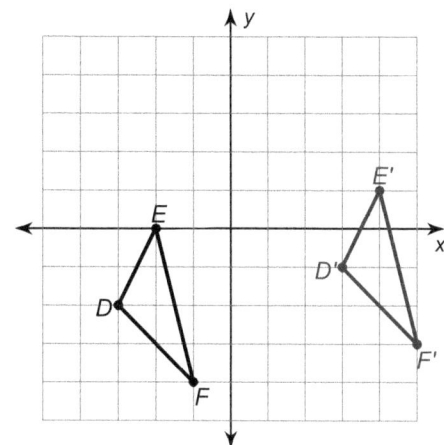

A) Translation: 5 units right and 3 units down
B) Translation: 2 units left and 2 units up
C) Ttranslation: 6 units right and 1 unit up
D) Translation: 7 units left and 1 unit down

6)

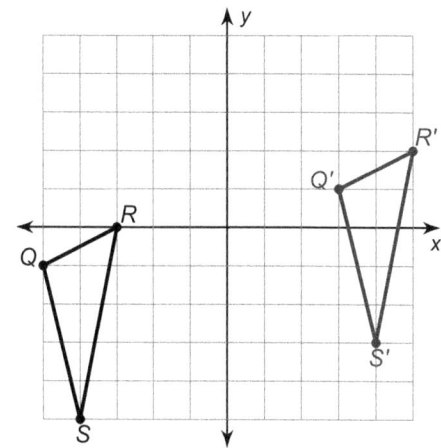

A) Translation: 8 units right and 2 units up
B) Translation: 8 units left
C) Translation: 5 units down
D) Translation: 2 units right and 8 units down

7)

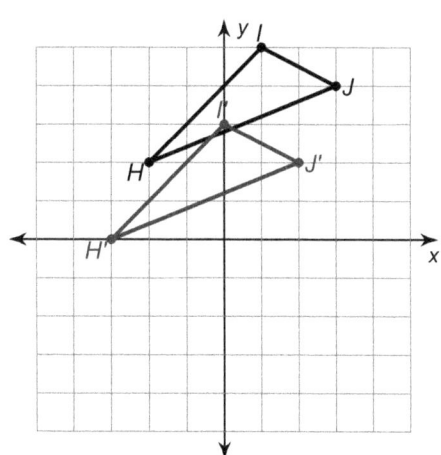

A) Translation: 4 units right and 1 unit up
B) Translation: 3 units up
C) Translation: 1 unit right
D) Translation: 1 unit left and 2 units down

Warm up 8

1. Evaluate $z\left(\dfrac{y-y}{6}\right) + 5$; where $y = -9$, and $z = -10$

   (A) 3   (B) 6

   (C) 8   (D) 5

2. Evaluate $|-9a - 2| = 2$, and find the sum of the solutions.

   (A) 7   (B) $-\dfrac{4}{9}$

   (C) 4   (D) -2

3. If $\left(\dfrac{1}{8}\right)^{-a} = 2^{-6}$ then find the value of a

   (A) 4   (B) -1

   (C) $\dfrac{1}{3}$ (D) -2

4. Find $(x + y)^2$ if $17x + 16y = -64$ and $5x - 16y = -288$

   (A) 9   (B) -9

   (C) 18   (D) 21

5. A straight line of slope $-\dfrac{1}{8}$ is passing through the points $A(8, -2)$ and $B(x, -1)$. Find the value of x.

   (A) -3   (B) 0

   (C) 1   (D) 3

6. Find the equation of a straight line passing through $H(-2, 2)$ and perpendicular to $y = -\dfrac{1}{2}x - 4$

   (A) $y = -2x + 6$ (B) $y = 2x + 6$

   (C) $y = 6x - 2$ (D) $y = -3x - 2$

7. Solve the below inequality

$$18k + 14 \leq 20k - 8 \leq 19k + 11$$

(A) $k \geq -2$ (B) $11 \leq k \leq 19$

(C) { All real numbers. } (D) $-11 \leq k \leq -1$

8. A picture is 3 cm wide and 2 cm tall. If it is enlarged to a width of 18 cm, then what is the new height ?

(A) 12 cm (B) 108 cm

(C) 3 cm (D) 11 cm

9. The original price of a music CD is $22.50 and discount offered on it is 30%. What is the selling price of the music CD ?

(A) $6.75 (B) $21.38

(C) $15.75 (D) $29.25

Practice test 8

1. Jack sells an Nintendo game at $75.87 for a 55% profit. Find the original price of the Nintendo game.

   (A) $48.95
   (B) $45.95
   (C) $29.37
   (D) $34.42

2. Dora pays $63.55 for a dress at the billing counter. If the store charges a 6% tax what is the price of the dress before tax ?

   (A) $43.55
   (B) $56.95
   (C) $56.35
   (D) $59.95

3. David added 81.7 instead of 66 in his calculations. Find the percentage change in his calculation.

   (A) 192.2 % increase
   (B) 18.02 % increase
   (C) 19.22 % increase
   (D) 195.9% decrease

4. Siya gave a personal loan to Tina an amount of $10,000 at 10% interest. If Tina repays it after 2 years, how much amount did Siya received ?

   (A) $12,100.00
   (B) $11,000.00
   (C) $2,000.00
   (D) $12,000.00

5. Barbara gave a personal loan to Frank an amount of $58,700 at 9% interest compounded annually. If Frank repays it after 3 years, how much amount will Barbara receive ?

   (A) $74,549.00
   (B) $76,009.36
   (C) $76,018.20
   (D) $76,024.21

6. There are 12 animals in the barn. Some are ducks and some are horses. There are 34 legs in all. How many of each animal are there ? How many horses are there in the barn ?

   (A) 4 horses
   (B) 3 horses
   (C) 5 horses
   (D) 2 horses

Practice Test - 8

7. Find the missing terms in the below sequence

..., 32, ___, ___, ___, 24, ...

(A) 35, 37, 39 (B) 37, 39, 41

(C) 30, 28, 26 (D) 34, 36, 38

8. Find the missing terms in the below sequence

..., -4, ___, -64, ...

(A) $-\dfrac{16}{3}$ (B) 24

(C) -8 (D) -16

9. If $P(A) = \dfrac{2}{5}$ $P(A \text{ and } B) = \dfrac{6}{25}$ then $P(B) = $?

(A) $\dfrac{3}{5}$ (B) $\dfrac{1}{2}$

(C) $\dfrac{3}{40}$ (D) $\dfrac{7}{20}$

10. A metallurgist needs to make 14 oz. of an alloy containing 83% silver. He is going to melt and combine one metal that is 89% silver with another metal that is 75% silver. How much of each should he use?

(A) 8 oz. of 89% silver, 6 oz. of 75% silver

(B) 4 oz. of 89% silver, 10 oz. of 75% silver

(C) 5 oz. of 89% silver, 9 oz. of 75% silver

(D) 10 oz. of 89% silver, 4 oz. of 75% silver

11. A diesel train left Miami and traveled south at an average speed of 10 mph. Sometime later a freight train left traveling in the opposite direction with an average speed of 50 mph. After the diesel train had traveled for 16 hours the trains were 710 mi. apart. How long did the freight train travel ?

(A) 14 hours (B) 11 hours

(C) 5 hours (D) 6 hours

12. Working alone, it takes Brenda 15 hours to tar a roof. Mei can tar the same roof in 10 hours. How long would it take them if they worked together?

    (A) 4.83 hours (B) 6 hours

    (C) 5.43 hours (D) 7.22 hours

13. There are five nickels and six dimes in a coin box. Rik randomly picks a coin out of the coin box and then returns it to the coin box. He randomly picks another coin. Find the probability of picking the first coin a nickel and the second coin a dime.

    (A) $\dfrac{1}{1024}$ (B) $\dfrac{2}{11}$

    (C) $\dfrac{1}{8}$ (D) $\dfrac{30}{121}$

14. The volume of a sphere is 7238.2 cu.ft. Find the diameter of the sphere. (Round your answer to the nearest tenth)

    (A) 24 ft (B) 15.2 ft

    (C) 12.8 ft (D) 30.8 ft

15. The ratio of the radii of two cylinders A and B is 3 : 4 respectively. The ratio of the height of cylinders A and B is 5 : 3 respectively. What is the ratio of the volume of cylinders A to B ?

    (A) 11 : 17 (B) 8 : 13

    (C) 21 : 19 (D) 15 : 16

16. How many 6 letter words with or without meaning can be formed from the letters of the word **PRECAUTION**, if repetition of the letters is not allowed ?

    (A) 380 (B) 2,840

    (C) 151,200 (D) 56,600

17. Find the average of first 40 natural numbers ?

    (A) 20.5 (B) 82

    (C) 41.5 (D) 22.4

18. A train X of length 180 m crosses another train Y traveling in opposite direction in 15 seconds. Train X crosses a pole in 20 seconds and the length of train X is half of length of train Y. Find the speed of train Y ?

   (A) 27 mi/s (B) 33 mi/s
   (C) 39 mi/s (D) 21 mi/s

19. Two machines X and Y takes 2 inputs and produces one output which is greatest of all inputs received. Machine Z takes the outputs of machines X and Y as inputs. Machine Z multiplies its inputs and produces it as output. Based on the given information and the below figure find the missing value ?

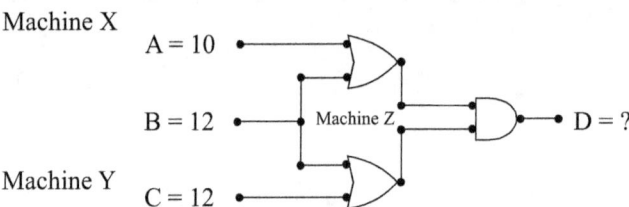

   (A) 144 (B) 12
   (C) 120 (D) 22

20. Ron melts a metal alloy weighing 8 mg containing 80% of iron and mixes it with another alloy weighing 1 mg containing 44 % iron. What percent of the resulting alloy is iron ?

   (A) 76 % (B) 63 %
   (C) 49 % (D) 67 %

21. Jason gives an amount of $16,900 as a personal loan to his friend Cathy at a interest rate of 14% compounded annually for 3 years. How much money does Cathy needs to pay at the end of the loan term ?

   (A) $ 23,998.55 (B) $ 25,031.05
   (C) $ 25,038.09 (D) $ 25,020.46

22. When you reverse the digits of a two digit number , we increase its value by 72. If the sum of its digits is 10 , find the two digit number ?

   (A) 19 (B) 26
   (C) 25 (D) 28

Practice Test - 8

23. The sum of a rational number and its reciprocal is $\frac{13}{6}$. Find the number ?

(A) $1\frac{5}{2}$ (B) $\frac{3}{2}$

(C) $\frac{1}{2}$ (D) $\frac{1}{3}$

24. The diagonal of a cube is $6\sqrt{3}$ inches. Find the volume of the cube ?

(A) 9 in^3 (B) 18 in^3

(C) 216 in^3 (D) 27 in^3

25. Find the missing terms in the below number series

..., 29, ___, ___, ___, ___, ___, 83, ...

(A) 50, 57, 64, 71, 78 (B) 38, 47, 56, 65, 74

(C) 43, 50, 57, 64, 71 (D) 36, 43, 50, 57, 64

26. Write the explicit for the below number series

$-\frac{15}{4}, -3, -\frac{5}{2}, -\frac{15}{7}, -\frac{15}{8}, ...$

27. A magic bag contains dimes, half dollar and dollar coins in the ratio of 5 : 4 : 3. If there are $22 worth of coins in the bag, find the total number of half dollar coins ?

(A) 14 (B) 16

(C) 12 (D) 18

28. Johnson spends 60% of his income for his education. He spends 30% of remaining on food and 50 % of the remaining for transportation and rent. If he saves $2,800 after all the expenses, find the amount he spends on his education ?

(A) $ 12,000 (B) $ 20,000

(C) $ 24,000 (D) $ 28,000

Warm up 9

1. Evaluate $\left(\dfrac{h}{6}\right)^2 + j - j$; where $h = -6$, and $j = 7$

   (A) 9   (B) 1

   (C) 7   (D) −5

2. Evaluate $|10 - 5b| = 20$, and find the sum of the solutions.

   (A) 4   (B) −2

   (C) −6   (D) −1

3. If $16^{-p-1} = 2^3$, find the value of p ?

   (A) $-\dfrac{7}{4}$ (B) −1

   (C) $\dfrac{7}{4}$ (D) 7

4. Find $2(x + y)$ if $32x + 5y = 80$ and $2x - 5y = 90$

   (A) -10   (B) 22

   (C) -22   (D) -11

5. A straight line of slope $-\dfrac{3}{4}$ is passing through the points $A(0, 2)$ and $B(-8, y)$. Find the value of x.

   (A) -1   (B) 0

   (C) 8   (D) -2

6. Find the equation of a straight line passing through $I(-5, -3)$ and perpendicular to $y = -\dfrac{5}{3}x - 4$

   (A) $y = \dfrac{3}{5}$ (B) $y = -\dfrac{3}{5}x + \dfrac{3}{5}$

   (C) $y = \dfrac{3}{5}x$ (D) $y = \dfrac{3}{5}x + \dfrac{3}{5}$

Practice Test - 9

7. Simplify the below inequality

$$9n - 13 \geq 8n + 12 \text{ or } 10n - 5 < -15n - 5$$

(A)   n > 9   (B)   n > 9 or n ≤ 1

(C)   n ≥ 25 or n < 0   (D)   n > 0 or n < −2

8. A picture is 2 mm wide and 3 mm tall. If it is enlarged to a height of 6 mm, find the new width ?

(A)   4 mm   (B)   12 mm

(C)   1 mm   (D)   3 mm

9. A new X - Box game is costing $99.50. Sofia bought the X - Box game at the discounted price of 53%. How much did Sofia paid ?

(A) $94.52   (B) $152.24

(C) $52.73   (D) $46.77

Practice test 9

1. Rachel sells a Lego set for $119.94 for a 20% profit. Find the original price of the Lego set?

   (A) $84.96   (B) $84.95

   (C) $99.95   (D) $79.96

2. Dora pays $37.05 for a bag at the billing counter. If the store charges a 6% tax what is the price of the bag before tax?

   (A) $29.71   (B) $32.85

   (C) $34.95   (D) $42.10

3. John added 43 instead of 85 in his calculations. Find the percentage change in his calculation.

   (A) 49.4% decrease   (B) 50.6% decrease

   (C) 97.7% increase   (D) 97.7% decrease

4. Lina took a loan of $310 at 15% interest for 2 years. Find the amount she needs to pay at the end of the loan term.

   (A) $ 356.50   (B) $ 403.00

   (C) $ 409.97   (D) $ 99.97

5. Alex took a loan of $50,000 at 4% interest compounded semi annually for 8 years. Find the amount he needs to pay at the end of the loan term.

   (A) $ 93,649.06   (B) $ 68,428.45

   (C) $ 66,000.00   (D) $ 68,639.29

6. Tracy bought 11 games for a total of $293. Game A cost $25 and Game B cost $28. How many number of game A's did she buy?

   (A) 5 Game A   (B) 8 Game A

   (C) 7 Game A   (D) 9 Game A

7. Find the number of unique permutations of the letters in the word **WATERMELON** ?

   (A) 1,814,400   (B) 1,612,200

   (C) 9,07,200   (D) 2,02,500

8. Find the missing terms in the below sequence

   ..., 24, ___, ___, ___, 52, ...

   (A) 30, 36, 42   (B) 17, 10, 3

   (C) 31, 38, 45   (D) 15, 8, 1

9. Find the missing terms in the below sequence ?

   ..., 3, ___, 108, ...

   (A) $\frac{2}{5}$ (B) 10

   (C) 15   (D) 18

10. If $P(A) = \frac{1}{4}$ $P(A \text{ and } B) = \frac{7}{80}$ then $P(B) = ?$

    (A) $\frac{7}{10}$ (B) $\frac{1}{5}$

    (C) $\frac{9}{20}$ (D) $\frac{7}{20}$

11. A metallurgist needs to make 16 oz. of an alloy containing 70% platinum. He is going to melt and combine one metal that is 85% platinum with another metal that is 55% platinum. How much of each should he use?

    (A) 8 oz. of 85% platinum, 8 oz. of 55% platinum

    (B) 5 oz. of 85% platinum, 11 oz. of 55% platinum

    (C) 13 oz. of 85% platinum, 3 oz. of 55% platinum

    (D) 11 oz. of 85% platinum, 5 oz. of 55% platinum

12. Find the number of terms in the below arithmetic series

$$a_1 = 7, a_n = 239, s_n = 3690$$

(A) 26 (B) 28

(C) 30 (D) 29

13. Evaluate the below series

$$\sum_{n=1}^{8} 4^{n-1}$$

(A) 17648 (B) 24955

(C) 23542 (D) 21845

14. Rik flips a coin five times. Find the probability of the coin landing heads-up the first two times and then lands tails-up the remaining three times.

(A) $\dfrac{64}{2197}$ (B) $\dfrac{30}{121}$

(C) $\dfrac{1}{32}$ (D) $\dfrac{28}{55}$

15. Sam's sock box has 6 white socks, 2 brown socks and 2 black socks. Find the probability of him randomly choosing two socks and get a matching pair of black socks ?

(A) $\dfrac{1}{17}$ (B) $\dfrac{1}{45}$

(C) $\dfrac{8}{33}$ (D) $\dfrac{2}{55}$

16. Find the missing terms in the below sequence

..., -19, ___, ___, ___, ___, ___, -49, ...

(A) -23, -26, -29, -32, -35 (B) 24, -29, -34, 39, 44

(C) -25, -30, -35, -40, -45 (D) -24, -29, -34, -39, -44

17. The perimeter of a rectangle is x feet and the circumference of a circle is 8 feet more than the perimeter of the rectangle. The ratio of a radius of a circle to the length of the rectangle is 7 : 12. The length and width of the rectangle are in the ratio of 3 : 2. Find the area of the rectangle ?

(A) 229 sq.ft (B) 216 sq.ft

(C) 284 sq.ft (D) 384 sq.ft

Quant Q TJHSST

Practice Test - 9

18. If the compound interest on a certain sum for 2 years at 10% per annum is $6300. Find the simple interest for the same amount of investment.

   (A) $ 5800
   (B) $ 6000
   (C) $ 6100
   (D) $ 5200

19. A and B together can complete the work in 12 days. A is 50 % more faster than B. In how many days the work can be completed if they both work on alternate days starting with A

   (A) 24 days
   (B) 20 days
   (C) 16 days
   (D) 18 days

20. If the numerator of a fraction is multiplied by $\frac{5}{7}$ and the denominator of the fraction is decreased by 20% then the fraction equals $\frac{5}{8}$. Find the sum of $\frac{3}{7}$ th percent of fraction and 25% of the fraction.

   (A) $\frac{19}{40}$
   (B) $\frac{15}{40}$
   (C) $\frac{17}{30}$
   (D) $\frac{13}{27}$

21. The sum of a digits of a certain a two digit number is 11. If the digits are reversed the value of the number increases by 9. What is the number ?

   (A) 56
   (B) 65
   (C) 47
   (D) 83

22. Find the average of first 20 multiples of 7 ?

   (A) 62.5
   (B) 73
   (C) 73.5
   (D) 92.4

23. If three numbers X, Y and Z are added in pairs then the sum equals to 10, 19, and 21. Find the numbers.

   (A) 6, 4 and 15
   (B) 6, 4 and 9
   (C) 15, 10 and 2
   (D) 6, 7 and 8

24. How many terms are there in the below arithmetic series ?

   $$\sum_{i=1}^{n} (8i - 7) = 855$$

   (A) 17
   (B) 15
   (C) 11
   (D) 16

25. Two machines X and Y takes 2 inputs and produces one output which is greatest of all inputs received. Machine Z takes the outputs of machines X and Y as inputs. Machine Z multiplies its inputs and produces it as output. Based on the given information and the below figure find the missing value?

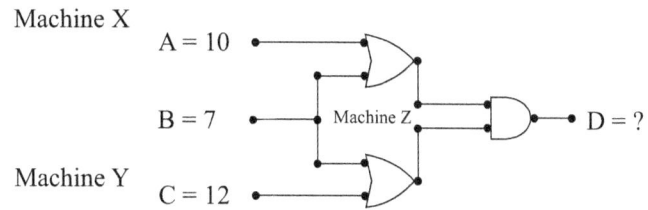

(A) 112 (B) 22

(C) 120 (D) 49

26. The volume of a cone with a height 11 cm is 103.7 cu.cm. Find the diameter of the cone?
(Round the answer to the nearest tenth)

(A) 3 cm (B) 6 cm

(C) 5 cm (D) 9.4 cm

27. A cylinder with a diameter of 10 yards has a volume of 235.6 cu.yd. Find the height of the cylinder?

(A) 8 yd (B) 5.2 yd

(C) 6.6 yd (D) 3 yd

28. Write the explicit formula for the below number series

10, 30, 50, 70, 90, ...

Warm up 10

1. Evaluate $x + y + x - |y|$; where $x = -10$, and $y = -5$

   (A) −30 (B) −28

   (C) −33 (D) −24

2. Evaluate $|2m - 1| = 13$, and find the sum of the solutions.

   (A) 20 (B) $\frac{7}{4}$

   (C) -4 (D) 1

3. If $36^{-3n-3} = 216^{-2n+3}$, find the value of n

   (A) No solution (B) $-\frac{3}{4}$

   (C) $-\frac{4}{5}$ (D) All real numbers

4. Solve for (x, y) from the below system of equations

   $$3x + 7y = -7$$
   $$3x + 7y = 112$$

   (A) (11, 9) (B) (9, 11)

   (C) (−9, 8) (D) No solution

5. A straight line of slope $\frac{1}{2}$ is passing through the points $A(x, -3)$ and $B(-9, -5)$. Find the value of x.

   (A) 7 (B) -8

   (C) -5 (D) -2

6. Find the equation of a straight line passing through J(2, 1) and perpendicular to $y = 2x - 1$

   (A) $y = -\frac{1}{2}x + 2$ (B) $y = \frac{1}{2}x + \frac{1}{2}$

   (C) $y = 2x + \frac{1}{2}$ (D) $y = \frac{1}{2}x + 2$

7. Solve the below inequality

$$-12 - 7m < -8 - 9m \text{ or } -16 + 17m > 15m - 6$$

(A) m < 2 or m > 5   (B) m < 0

(C) m > 5   (D) m < 2

8. Evaluate $\dfrac{^{10}C_5}{7} + 4 = ?$

(A) 40   (B) 18

(C) 30   (D) 10

9. A printing paper supply box costs $200 with a discount of 54% on it. What is the selling price of the printing paper box ?

(A) $ 92   (B) $ 109

(C) $ 120   (D) $ 108

Practice test 10

1. Alex purchased a Music CD and sold it for $18.44. He makes a profit of 23% on the sale. Find the purchase price of the Music CD.

   (A) $15.45   (B) $16.49

   (C) $23.45   (D) $14.99

2. Noor pays $20.28 for a shirt at the billing counter. If the store charges a 4% tax what is the price of the shirt before tax ?

   (A) $25.28   (B) $19.50

   (C) $21.90   (D) $18.72

3. John added 32 instead of 19 in his calculations. Find the percentage change in his calculation.

   (A) 68.4% increase   (B) 13% decrease

   (C) 13% increase   (D) 168.4% increase

4. Misha took a loan of $43,000 at 7% interest for 2 years. Find the amount she needs to pay at the end of the loan term.

   (A) $3,010.00   (B) $49,230.70

   (C) $49,020.00   (D) $6,230.70

5. Skylar took a loan of $46,000 at 8% interest compounded semi annually for 2 years. Find the amount she needs to pay at the end of the loan term.

   (A) $53,813.49   (B) $53,360.00

   (C) $62,582.49   (D) $53,654.40

6. Tracy bought 11 games for a total of $293. Game A cost $25 and Game B cost $28. How many number of game B's did she buy ?

   (A) 6 Game B   (B) 3 Game B

   (C) 4 Game B   (D) 2 Game B

7. Find the missing terms in the below sequence

   ..., 31, ___, ___, ___, 11, ...

   (A) 26, 21, 16   (B) 23, 19, 15

   (C) 27, 23, 19   (D) 25, 23, 21

Practice Test - 10

8. Find the missing terms in the below sequence

   ..., 3, ___, ___, ___, ___, ___, 63, ...

   (A) -9, -21, -33, -45, -57
   (B) 15, 27, 39, 51, 63
   (C) 13, 23, 33, 43, 53
   (D) -7, -17, -27, -37, -47

9. Find the missing terms in the below sequence

   ..., 2, ___, 72, ...

   (A) 1
   (B) 7
   (C) 6
   (D) 12

10. Write the explicit formula for the below number series

    -6, -2, 2, 6, 10, ...

11. If $P(A) = \dfrac{1}{4}$ $P(B|A) = \dfrac{3}{20}$ then $P(A \text{ and } B) = $?

    (A) $\dfrac{3}{80}$
    (B) $\dfrac{11}{25}$
    (C) $\dfrac{7}{20}$
    (D) $\dfrac{13}{25}$

12. An acid solution was made by mixing 4 fl. oz. of a 58% acid solution and 2 fl. oz. of a 28% acid solution. What is the concentration of the mixture ?

    (A) 41 %
    (B) 10 %
    (C) 55 %
    (D) 48 %

13. Train A left Virginia and traveled east at an average speed of 60 mph. Train B leaves sometime later traveling in the opposite direction with an average speed of 50 mph. After train A had traveled for six hours the trains were 560 mi. apart. Find the number of hours travelled by train B ?

　(A)　1 hour　　　　　　　　　　　　　(B)　2 hours

　(C)　5 hours　　　　　　　　　　　　　(D)　4 hours

14. Bill can tar a roof in 13 hours. Mike can tar the same roof in 9 hours. How long would it take them if they worked together ?

　(A)　5.03 hours　　　　　　　　　　　(B)　4.9 hours

　(C)　6.47 hours　　　　　　　　　　　(D)　5.32 hours

15. A basket contains five apples and seven peaches. Rak randomly selects a piece of fruit and then returns it to the basket and picks another fruit. Find the probability that she picks apples in both trials.

　(A)　$\dfrac{1}{4}$　　　　　　　　　　　　　(B)　$\dfrac{25}{144}$

　(C)　$\dfrac{8}{33}$　　　　　　　　　　　　(D)　$\dfrac{1}{8}$

16. Find the number of terms in the arithmetic series

$$a_1 = 10 \, , \, a_n = 100 \, , \, s_n = 550$$

　(A)　11　　　　　　　　　　　　　　　(B)　9

　(C)　12　　　　　　　　　　　　　　　(D)　10

17. Rak has 5 blue dresses , 4 green dresses and 5 red dresses in her closet. She randomly selects a different dress each day. Find the probability of her picking blue dress on Monday , green dress on Tuesday and red dress on Wednesday.

　(A)　$\dfrac{25}{546}$　　　　　　　　　　　(B)　$\dfrac{7}{26}$

　(C)　$\dfrac{3}{110}$　　　　　　　　　　　(D)　$\dfrac{1}{22}$

Quant Q TJHSST

Practice Test - 10

18. The surface area of a cube is 1734 sq.in. Find the volume of the cube ?

    (A) 4903 in³   (B) 4612 in³

    (C) 4725 in³   (D) 4913 in³

19. The average of four consecutive even numbers is 27. Find the largest of these numbers.

    (A) 30   (B) 24

    (C) 28   (D) 32

20. A bag is sold at certain price. by selling it by $\frac{7}{11}$ th of the price Dan loses 30 %. Find the percentage of gain from the original selling price.

    (A) 20 %   (B) 45 %

    (C) 10 %   (D) 30 %

21. A stream is running at 5 mi/hr. Ron travels on a boat at 25 miles upstream and back in 3 hours 45 mins. Find the speed of Ron's boat.

    (A) 10 mi/hr   (B) 20 mi/hr

    (C) 15 mi/hr   (D) 30 mi/hr

22. In how many different ways the group of 2 men and 3 women can be formed out of 5 men and 5 women ?

    (A) 720   (B) 60

    (C) 120   (D) 100

23. The length and breadth of the rectangular piece of land are in the ratio of 3 : 2. Jacob spends $5,700 for making a fence at the rate of $9.50 per foot. Find the area of rectangular field.

    (A) 19,500 sq.ft   (B) 21,600 sq.ft

    (C) 28,400 sq.ft   (D) None of these

24. The ratio of curved surface area to the total surface area of a right circular cylinder is 3 : 8. Find the volume of the cylinder if the area of the base is 3850 sq.m.

    (A) 82,260 sq.m   (B) 26,480 sq.m

    (C) 80,850 sq.m   (D) 72,540 sq.m

25. Two numbers are in the ratio of 11 : 23. If their GCF is 6 find the average of the numbers.

   (A)   102   (B)   204

   (C)   68   (D)   105

26. Two machines X and Y takes 2 inputs and produces one output which is greatest of all. Machine Z takes the outputs of machines X and Y as inputs. Machine Z multiplies its inputs and produces it as output. Based on the given information and the below figure find the missing value ?

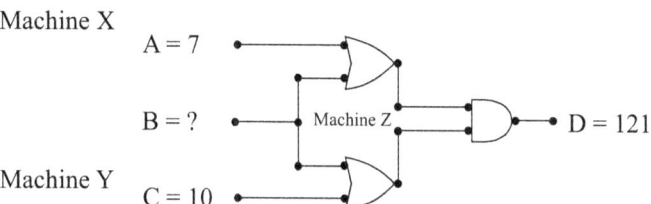

   (A)   11   (B)   101

   (C)   100   (D)   55

27. Find the number of unique permutations of the letters of the word **MICROSCOPE** ?

   (A)   907,200   (B)   454,000

   (C)   964,500   (D)   108,058

28. A rectangular prism with 17 cm and 10 cm along the base has a volume of 1700 cu.cm. Find the height of the rectangular prism ?

   (A)   5 cm   (B)   25 cm

   (C)   16 cm   (D)   10 cm

Warm up 11

1. Evaluate $y + x^2(x + 4)$; where $x = -2$, and $y = -4$

   (A)   −3   (B)   11

   (C)   1   (D)   4

2. Find the solutions of $|8x + 2| = 78$

   (A) $\left\{\dfrac{19}{2}\right\}$ (B) $\{-3, -7\}$

   (C) $\{4, -8\}$ (D) $\left\{\dfrac{19}{2}, -10\right\}$

3. Evaluate $27^{-v} = \left(\dfrac{1}{3}\right)^{3v}$, Find the value of "v".

   (A)   All real numbers   (B)   −10

   (C)   1   (D) $-\dfrac{11}{9}$

4. Find (x, y) by solving the below equations

   $$x + 3y = 54$$
   $$23x - 15y = 150$$

   (A)   (13, 15)   (B)   No solution

   (C)   (15, −13)   (D)   (15, 13)

5. A straight line of slope -3 passing through $A(x, -5)$ and $B(4, -8)$. Find the value of x

   (A)   7   (B)   1

   (C)   3   (D)   9

6. A box of Apples costs $14.50. Food mart is offering 25% discount on it today. Find the selling price of the box of Apples.

   (A) $10.88   (B) $13.77

   (C) $15.95   (D) $3.62

7. Find the equation of a straight line passing through K(−4, 4) and perpendicular to y = 4x + 3

   (A) $y = -\frac{3}{4}x + 3$ (B) $y = -\frac{5}{4}x + 3$

   (C) $y = \frac{5}{4}x + 3$ (D) $y = -\frac{1}{4}x + 3$

8. Solve for "n" from the below inequalities

   $3n + 11 \leq 3 + 4n$ and $4n - 19 < 7n + 14$

   (A) All real numbers (B) $n > -11$

   (C) $-3 < n < 3$ (D) $n \geq 8$

9. Rosy reduced the size of a painting to a width of 3 cm. What is the new height if it was originally 3 cm tall and 9 cm wide?

   (A) 1 cm (B) 9 cm

   (C) 2 cm (D) 3 cm

Quant Q
TJHSST

Practice Test - 11

Practice test 11

1. Andy sold a Harry potter Book for $5.45 and makes a 10% profit. Find the cost price of the Harry potter Book.

   (A) $4.21   (B) $4.70

   (C) $6.50   (D) $4.95

2. If $1.50 is the original price of a magazine, calculate the selling price of the magazine, when 2% tax is added to it.

   (A) $1.53   (B) $1.65

   (C) $1.72   (D) $0.03

3. John added 37 instead of 24 in his calculations. Find the percentage change in his calculation.

   (A) 54.2% decrease   (B) 54.2% increase

   (C) 7% decrease   (D) 35.1% decrease

4. Tony gave a personal loan to Chug an amount of $1,100 at 2% interest. If Chug repays it after 3 years, How much amount did Tony received ?

   (A) $ 22.00   (B) $ 1,122.00

   (C) $ 66.00   (D) $ 1,166.00

5. Lucas gave a personal loan to Logan an amount of $4,000 at 2% interest, compounded semi annually. If Logan repays it after 3 years, how much amount will Lucas receive ?

   (A) $4,121.20   (B) $4,244.83

   (C) $4,240.00   (D) $4,246.08

6. Jack spent $610 on story books. Fiction books costs $50 and non-fiction books costs $60. He bought a total of 11 story books. How many number of fiction books did he buy ?

   (A) 5 Fiction books   (B) 9 Fiction books

   (C) 8 Fiction books   (D) 8 Fiction books

Practice Test - 11

7. Find the missing terms in the below sequence

..., 11, ___, ___, ___, ___, ___, 611, ...

(A) 309, 409, 509, 609, 709

(B) 211, 311, 411, 511, 611

(C) 111, 211, 311, 411, 511

(D) 209, 309, 409, 509, 609

8. Find the missing terms in the below sequence

..., -1, ___, ___, -64, ...

(A) 3, 12

(B) -4, -16

(C) -1, -4

(D) 7, 28

9. Write the explicit formula for the below number series

13.7, 15, 16.3, 17.6, 18.9, ...

10. If $P(A) = \frac{1}{5}$ $P(A \text{ and } B) = \frac{1}{20}$ then $P(B) = $?

(A) $\frac{7}{20}$

(B) $\frac{1}{4}$

(C) $\frac{9}{50}$

(D) $\frac{3}{4}$

11. If $P(B) = \frac{2}{5}$ $P(A \text{ and } B) = \frac{3}{10}$ then $P(A|B) = $?

(A) $\frac{7}{20}$

(B) $\frac{7}{10}$

(C) $\frac{1}{2}$

(D) $\frac{3}{4}$

12. A boat travels at a speed of 22 mi/hr along with the water current. Traveling against the same current it travels at a speed of 4 mi/hr. Find the speed of the boat in still water.

   (A) 21 mi/hr   (B) 4 mi/hr

   (C) 13 mi/hr   (D) 9 mi/hr

13. An acid solution was made by mixing 4 fl.oz. of a 58% acid solution and 2 fl. oz. of a 28% acid solution. What is the concentration of the mixture?

   (A) 41 %   (B) 10 %

   (C) 55 %   (D) 48 %

14. Diana has eight nickels and eight dimes in her coin bag. Three times, she randomly picks a coin out of the bag, returns it back and mixes up the coins. Find the probability of her picking a nickel in the first try, a dime in the second try and nickel again in the last try.

   (A) $\dfrac{1}{8}$ (B) $\dfrac{1}{64}$

   (C) $\dfrac{2}{11}$ (D) $\dfrac{1}{22}$

15. Rik has thirteen shirts in his closet, four blue, five green, and four red. He randomly selects a different shirt each day. Find the probability of him wearing a blue shirt on Monday, Tuesday, and Wednesday.

   (A) $\dfrac{14}{33}$ (B) $\dfrac{20}{429}$

   (C) $\dfrac{2}{55}$ (D) $\dfrac{2}{143}$

16. Two pipes A and B can fill an empty tank in 12 hours and 15 hours respectively. If both the pipes are opened simultaneously, find the time take to fill $\dfrac{1}{6}$ th of the tank ?

   (A) $\dfrac{15}{2}$ hours   (B) $\dfrac{16}{3}$ hours

   (C) $\dfrac{10}{9}$ hours   (D) $\dfrac{29}{3}$ hours

17. Rectangular wood block of 6 cm X 12 cm X 15 cm is cut into exact number of equal cubes. Find the least possible number of cubes that can be cut from the wood block.

    (A) 40

    (B) 27

    (C) 34

    (D) 46

18. Sum of a two digit number is 12. If we reverse the digits, the value of the number increases by 36. Find the number.

    (A) 48

    (B) 58

    (C) 84

    (D) 66

19. The mean of 5 observations of a science experiment is 11. If the observations are $x, x+2, x+4, x+6, x+8$ then is mean of last 3 observations is

    (A) 11

    (B) 13

    (C) 15

    (D) 17

20. Two machines X and Y takes 2 inputs and produces one output which is greatest of all. Machine Z takes the outputs of machines X and Y as inputs. Machine Z multiplies its inputs and produces it as output. Based on the given information and the below figure find the missing value ?

    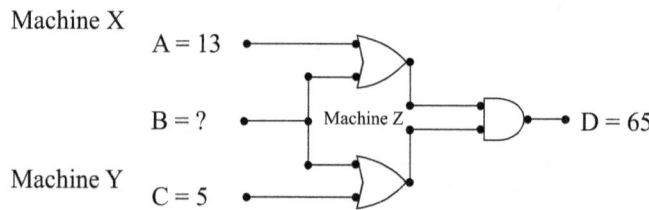

    (A) 3

    (B) 25

    (C) 9

    (D) 15

21. Find the missing terms in the below sequence

    ..., 2, ___, ___, ___, 22, ...

    (A) 9, 14, 19

    (B) 13, 18, 23

    (C) 7, 12, 17

    (D) 8, 13, 18

Quant Q TJHSST

Practice Test - 11

22. An arm of a ceiling fan measures 25 cm. When the fan is turned on find the area covered by the arm

   (A) 246.49 sq.cm   (B) 78.5 sq.cm

   (C) 1962.5 sq.cm   (D) 157 sq.cm

23. Find the area of the shaded portions in terms of ∏ where AB = 6 cm and BC = 8 cm from the below figure.

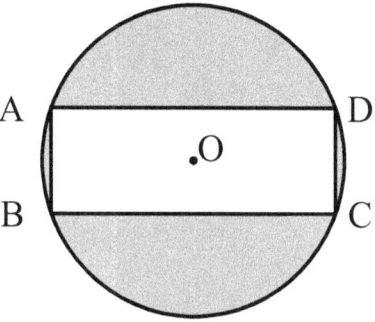

   (A) 25 ∏ - 72   (B) 25 ∏ - 48

   (C) 25 ∏ - 8   (D) 100 ∏ - 48

24. Elsa and Ila are selling big and small art pieces. Elsa sold nine small and six big art pieces for a total of $213. Ila sold six small and one big art pieces for a total of $94. What is the cost of each of big and small art pieces ?

   (A) bid art piece = $16 , small art piece = $13
   (B) bid art piece = $13 , small art piece = $10
   (C) bid art piece = $16 , small art piece = $23
   (D) bid art piece = $10 , small art piece = $15

25. The sum of the digits of certain two digit number equals to 9. When you reverse the digits the value is decreased by 45. Find the number ?

   (A) 79   (B) 87

   (C) 72   (D) 48

26. Find the number of unique permutations of the letters of the word **THURSDAY** ?

   (A) 40,320   (B) 45,280

   (C) 25,130   (D) 40,960

27. Find $f(7,3)$

$$f(x,y) = \begin{cases} f(x-3, y+2) & \text{when } x > y \\ f(x+y, y+4) & x < y \\ x + \dfrac{y}{x} + xy & x = y \end{cases}$$

(A) 64 (B) 85
(C) 56 (D) 91

28. Find C(5) if C(x) = C(x - 2) + x ans we know that C(1) = a , C(0) = 0

(A) a - 8 (B) a + 8
(C) a + 3 (D) a - 3

Warm up 12

1. Evaluate $\dfrac{(y-y)^2 + x}{2}$; where x = 10, and y = 8

   (A) −10 (B) −4
   (C) 11   (D) 5

2. Evaluate |3v + 8| = 2 , and find the sum of the solutions.

   (A) $-\dfrac{14}{5}$ (B) $-\dfrac{16}{3}$
   (C) $\dfrac{10}{3}$ (D) $\dfrac{6}{7}$

3. If $6^{3-3n} = 36$, find the value of n

   (A) $\dfrac{1}{3}$ (B) $-\dfrac{4}{5}$
   (C) $\dfrac{7}{5}$ (D) −1

4. Find $(x + y)^3$ by solving the below equations.

   $2x + 11y = 33$
   $18x + 11y = -143$

   (A) -216   (B) 64
   (C) 36   (D) 225

5. A straight line of slope $-\dfrac{2}{7}$ passing through A(1, y) and B(−6, 3). Find the value of x

   (A) -7   (B) 2
   (C) 1   (D) -9

6. Find the equation of a straight line passing through L(−3, −4) and perpendicular to y = 4x − 3

   (A) $y = -\dfrac{19}{4}x - \dfrac{1}{4}$ (B) $y = \dfrac{1}{4}x - \dfrac{19}{4}$
   (C) $y = -\dfrac{19}{4}x + \dfrac{1}{4}$ (D) $y = -\dfrac{1}{4}x - \dfrac{19}{4}$

7. Solve for "a" from the below inequality

$$299 \leq 5 + 6(1 - 6a)$$

(A) a > 6 (B) a ≥ -8

(C) a ≤ -8 (D) a > 48

8. Evaluate $^{12}P_4 - 5 = ?$

(A) 430 (B) 440

(C) 450 (D) 520

9. A box of Oranges costs $14.50. Food mart is offering 7% discount on it today. Find the selling price of the box of Oranges?

(A) $16.45 (B) $13.49

(C) $16.00 (D) $1.02

Practice test 12

1. Anthony sells a box of pencils for $3.50 and makes a profit of 75% from each sale. Find the cost price of the box of pencils.

   (A) $ 2.00   (B) $ 0.75

   (C) $ 1.50   (D) $ 1.90

2. The original price of a refrigerator is $15,000.00 and tax added to it is 6%. What is the sale price of the refrigerator ?

   (A) $ 900.00   (B) $ 17,250.00

   (C) $ 14,100.00   (D) $ 15,900.00

3. John added 34 instead of 41 in his calculations. Find the percentage change in his calculation.

   (A) 7% decrease   (B) 17.1% increase

   (C) 17.1% decrease   (D) 120.6% decrease

4. Dora gave a personal loan to Noor an amount of $38,500 at 14% interest. If Noor repays it after 4 years, how much amount did Dora received ?

   (A) $60,060.00   (B) $65,024.97

   (C) $26,524.97   (D) $43,890.00

5. James gave a personal loan to Liam an amount of $55,900 at 13% interest. compounded annually. If Liam repays it after 4 years, how much amount will James receive ?

   (A) $84,968.00   (B) $91,143.47

   (C) $91,161.29   (D) $91,124.96

6. Jack spent $610 on story books. Fiction books costs $50 and non-fiction books costs $60. He bought a total of 11 story books. How many number of non fiction books did he buy ?

   (A) 6 non-fiction books   (B) 5 non-fiction books

   (C) 3 non-fiction books   (D) 2 non-fiction books

Practice Test - 12

7. Find the missing terms in the below sequence

   ..., 36, ___, ___, 336, ...

   (A) 135, 235 (B) 133, 233

   (C) 136, 236 (D) -64, -164

8. Find the missing terms in the below sequence

   ..., -16, ___, ___, ___, ___, ___, 14, ...

   (A) -1, 4, 9, 14, 19 (B) -2, 3, 8, 13, 18

   (C) -6, -1, 4, 9, 14 (D) -11, -6, -1, 4, 9

9. Find the missing terms in the below sequence

   ..., -3, ___, -27, ...

   (A) -2 (B) 1

   (C) -9 (D) -1

10. Write the explicit formula for the below number series

    -32, 168, 368, 568, 768, ...

11. If $P(B) = \dfrac{1}{5}$ $P(A \text{ and } B) = \dfrac{7}{100}$ then $P(A) = ?$

    (A) $\dfrac{13}{20}$ (B) $\dfrac{3}{4}$

    (C) $\dfrac{3}{40}$ (D) $\dfrac{7}{20}$

Quant Q TJHSST

Practice Test - 12

12. If $P(A) = \dfrac{9}{20}$ $P(B|A) = \dfrac{13}{20}$ then $P(A \text{ and } B) = ?$

    (A) $\dfrac{3}{20}$
    (B) $\dfrac{91}{400}$

    (C) $\dfrac{117}{400}$
    (D) $\dfrac{3}{5}$

13. If $P(B) = \dfrac{1}{4}$ $P(A \text{ and } B) = \dfrac{1}{20}$ then $P(A) = ?$

    (A) $\dfrac{2}{5}$
    (B) $\dfrac{1}{5}$

    (C) $\dfrac{1}{2}$
    (D) $\dfrac{7}{20}$

14. Henry left his home an hour before Lena. They drove in opposite directions. Lena drove at 75 m/hr for three hours. After this time they were 485 m apart. Find the speed of Henry.

    (A) 65 m/hr
    (B) 55 m/hr

    (C) 70 m/hr
    (D) 30 m/hr

15. Cathy flips a coin and then rolls a fair six-sided dice. Find the probability of the coin lands heads-up and the dice shows an even number.

    (A) $\dfrac{125}{1331}$
    (B) $\dfrac{1}{36}$

    (C) $\dfrac{4}{15}$
    (D) $\dfrac{1}{4}$

16. Mary's refrigerator has ten bottles of sports drinks: six lemon-lime flavored and four orange flavored. She randomly grabs a bottle and gives it to her friend Jade. Then, she randomly grabs another bottle for herself. Find the probability of Jade getting a lemon-lime and Mary getting an orange sports drinks?

    (A) $\dfrac{8}{33}$
    (B) $\dfrac{1}{22}$

    (C) $\dfrac{25}{546}$
    (D) $\dfrac{4}{15}$

17. A price of puzzle A is 10 % more than the price of puzzle B. Puzzle B is what percent less than puzzle A ?

   (A)   11.25 %   (B)   10.11 %

   (C)   8.23 %   (D)   9.09 %

18. A milk jar contains 92 litres of mixtures of milk and water in the ratio of 19 : 4. Mary uses 46 % of the mixture and adds 4 litres of water to the jar. Find the percentage of water in the new mixture ?

   (A)   24 %   (B)   27 %

   (C)   32 %   (D)   18 %

19. Rik drives 360 km from city A to city B in 5 hours and returns back in 4 hours. If m is the average speed of the entire trip then the average speed of the journey from city B to city A exceeds m by how much ?

   (A)   10 km/hr   (B)   16 km/hr

   (C)   12 km/hr   (D)   8 km/hr

20. The speed of a boat in still water is 3 times the speed of the stream. The boat takes 9 hours to travel 180 miles downstream. Find the ratio of time taken by the boat to travel 80 miles downstream to 50 miles upstream.

   (A)   5 : 6   (B)   4 : 5

   (C)   3 : 4   (D)   2 : 3

21. If two Pipes are open simultaneously the tank will be filled in 7.2 hours. Pipe A can fill the tank in 6 hours faster than pipe B. How many hours does it take for pipe B to fill the tank alone ?

   (A)   18 hours   (B)   12 hours

   (C)   9 hours   (D)   27 hours

22. A bag contains 5 red balls 4 blue balls and 6 green balls. Three balls are drawn at random and are not replaced. Find the probability that at least one ball is red in color ?

   (A) $\dfrac{21}{23}$ (B) $\dfrac{47}{91}$

   (C) $\dfrac{11}{67}$ (D) $\dfrac{67}{91}$

23. A square prism having 14 inches height has a volume of 2016 cu.in. Find the length of the edge ?

   (A)   12 in   (B)   15 in

   (C)   10 in   (D)   11 in

24. A triangle ABC with a height of 5.7 yards and has an area of 22.8 sq.yd. Find the length of its base ?

   (A)   8 yd   (B)   12 yd

   (C)   3 yd   (D)   18 yd

25. The sum of the digits of certain two digit number is 5. By reversing its digits the value of the number increases the value of the number increases by 9. Find the number ?

   (A)   32   (B)   23

   (C)   50   (D)   41

26. Two machines X and Y takes 2 inputs and produces one output which is greatest of all inputs received. Machine Z takes the outputs of machines X and Y as inputs. Machine Z multiplies its inputs and produces it as output. Based on the given information and the below figure find the missing value ?

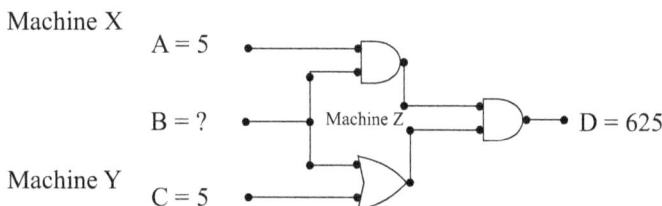

   (A)   10   (B)   200

   (C)   5   (D)   25

27. A cube of 15 cm is immersed completely in a rectangular prism containing water. If the dimensions of the rectangular vessel are 20 cm X 15 cm. Find the rise in water level ?

   (A)   11.25 cm   (B)   9.5 cm

   (C)   6.45 cm   (D)   10.5 cm

28. Find the number of unique permutations of the letters in the word **FEARLESSLY** ?

   (A)   253,300   (B)   545,290

   (C)   453,600   (D)   395,900

Warm up 13

1. Evaluate (p − (m − 2 + m − m)); where m = 1, and p = 3

   (A) 4   (B) 8

   (C) 13   (D) 7

2. Find the solutions of |−6n − 4| = 28

   (A) $\left\{8, -\dfrac{42}{5}\right\}$ (B) $\left\{8, -\dfrac{48}{5}\right\}$

   (C) $\left\{-\dfrac{16}{3}, 4\right\}$ (D) {8}

3. Evaluate $7^{-2k} = 49^{-k}$, Find the value of "k".

   (A) 9   (B) 5

   (C) −10   (D) All real numbers

4. Find (x , y) by solving the below equations

   $$y = -16$$
   $$3x + 2y = -14$$

   (A) (15, −16)   (B) (6, 16)

   (C) (15, 16)   (D) (6, −16)

5. A straight line of slope $\dfrac{17}{3}$ passing through A(9, 8) and B(6, y). Find the value of x

   (A) -8   (B) -9

   (C) -5   (D) 9

6. Stella purchased a Jar for $0.95. What is the selling price if the discount offered on the jar is 50% ?

   (A) $1.14   (B) $0.47

   (C) $1.42   (D) $0.48

7. Find the equation of a straight line passing through M(−4, −2) and perpendicular to $y = -\frac{2}{3}x - 5$

   (A) $y = \frac{3}{2}x + 4$
   (B) $y = 4x - \frac{3}{2}$
   (C) $y = -\frac{3}{2}x + 4$
   (D) $y = -4x - \frac{3}{2}$

8. Solve for "x" from the below inequalities

   $$13x + 13 \geq -3 + 14x \geq 7x + 18$$

   (A) $x \leq 0$
   (B) $-13 \leq x \leq -4$
   (C) $3 \leq x \leq 16$
   (D) $x \leq 10$

9. Bary enlarged the size of a photo to a width of 12 in. What is the new height if it was originally 3 in wide and 2 in tall?

   (A) 8 in
   (B) 48 in
   (C) 7 in
   (D) 1 in

Practice test 13

1. Brad sells a package Office supplies for $288.00 and makes a profit of 80%. Find the original price of the Office supplies package.

   (A) $160.00   (B) $132.00

   (C) $228.00   (D) $158.00

2. The original price of a refrigerator is $6.25 and tax added to it is 6%. What is the sale price of the refrigerator?

   (A) $7.50   (B) $0.38

   (C) $6.63   (D) $5.88

3. John added 31 instead of 53 in his calculations. Find the percentage change in his calculation.

   (A) 41.5% decrease   (B) 45.1% decrease

   (C) 51.4% increase   (D) 14.5% increase

4. Hanna took a loan of $19,300 at 5% interest for 2 years. Find the amount she needs to pay at the end of the loan term.

   (A) $21,230.00   (B) $1,978.25

   (C) $20,265.00   (D) $21,278.25

5. Grace took a loan of $405 at 5% interest compounded semi annually for 8 years. Find the amount she needs to pay at the end of the loan term.

   (A) $601.22   (B) $493.45

   (C) $567.00   (D) $598.37

6. Lucy bought 10 dresses for a total of $204. Formal dress cost $26 and casual dress cost $12. How many number of formal dress did she buy?

   (A) 6 formal dress   (B) 2 formal dress

   (C) 8 formal dress   (D) 7 formal dress

7. Find the missing terms in the below sequence

   ..., 7, ___, ___, ___, 87, ...

   (A) 27, 47, 67   (B) 25, 45, 65

   (C) 26, 46, 66   (D) 46, 66, 86

8. Find the missing terms in the below sequence

   ..., 23, ___, ___, ___, ___, ___, 83, ...

   (A) 21, 11, 1, -9, -19   (B) 41, 51, 61, 71, 81

   (C) 43, 53, 63, 73, 83   (D) 33, 43, 53, 63, 73

9. Find the missing terms in the below sequence

   ..., 4, ___, 64, ...

   (A) 64   (B) 4

   (C) 48   (D) 16

10. Write the explicit formula for the below number series

    $$\frac{1}{2}, \frac{1}{2}, \frac{3}{8}, \frac{1}{4}, \frac{5}{32} ...$$

11. If $P(A) = \frac{1}{5}$ $P(B) = \frac{7}{20}$ then $P(A \text{ and } B) = ?$

    (A) $\frac{27}{200}$ (B) $\frac{3}{10}$

    (C) $\frac{49}{200}$ (D) $\frac{7}{100}$

12. If $P(B) = \dfrac{3}{5}$ $P(A \text{ and } B) = \dfrac{9}{40}$ then $P(A|B) = ?$

   (A) $\dfrac{9}{100}$ (B) $\dfrac{1}{2}$

   (C) $\dfrac{5}{8}$ (D) $\dfrac{7}{40}$

13. If $P(\text{not } A) = \dfrac{1}{2}$ $P(B) = \dfrac{1}{4}$ then $P(A \text{ and } B) = ?$

   (A) $\dfrac{9}{20}$ (B) $\dfrac{1}{16}$

   (C) $\dfrac{1}{8}$ (D) $\dfrac{1}{2}$

14. Nidhi can clean an attic in 12 hours and Kristin can clean the same attic in 10 hours. How many hours will take them if they worked together ?

   (A) 5.45 hours   (B) 4.85 hours

   (C) 4.4 hours   (D) 4.75 hours

15. Mona rolls a fair six-sided dice twice. Find the probability of the first roll showing a two and the second roll showing a five.

   (A) $\dfrac{10}{273}$ (B) $\dfrac{36}{121}$

   (C) $\dfrac{1}{22}$ (D) $\dfrac{1}{36}$

16. Amanda buys a basket of fruits containing eight apples and four peaches. She randomly selects one piece of fruit and eats it. Then she randomly selects another piece of fruit. Find the probability of her selecting apples both times.

   (A) $\dfrac{14}{33}$ (B) $\dfrac{16}{49}$

   (C) $\dfrac{1}{22}$ (D) $\dfrac{25}{546}$

Quant Q TJHSST

Practice Test - 13

17. Manu was curious to know how many times digit 8 is repeated from 1 through 100 ?

   (A) 10 (B) 19

   (C) 30 (D) 8

18. A minute cell start to divide at 6AM into 6 parts for every 3 minutes. At 6.30. How many cells are formed ?

   (A) 6^9 (B) 6^5

   (C) 6^{10} (D) 6^6

19. If R#P = 70 - RP what is the value of

   1#(5#2)

   (A) 25 (B) 60

   (C) 30 (D) 10

20. Working together, Jack and Jill can sweep a porch in 5.32 minutes. Jill takes 9 minutes if he cleans by himself. In how many minutes can Jack clean the room by himself ?

   (A) 13.01 hours (B) 11.11 hours

   (C) 12.03 hours (D) 12.95 hours

21. Sam got twelve chips boxes and eight candy boxes from store A for $128. Next day he went to store B and bought ten chips boxes and four candy boxes for $80. What is the price for a box of chips and box of candy ?

   (A) Chips box = $4 , Candy box = $10
   (B) Chips box = $14 , Candy box = $11
   (C) Chips box = $10 , Candy box = $4
   (D) Chips box = $4 , Candy box = $15

22. Sam charges $15 for every 50 Sq ft of the wall to paint. How much will it cost for 8 walls of 300 sq ft each ?

   (A) $720 (B) $90

   (C) $120 (D) $180

23. A rectangular garden of length 12 yd has an area of 108 sq.yd. Find the width of the garden ?

   (A) 12 yd (B) 6 yd

   (C) 9 yd (D) 18 yd

24. Two machines X and Y takes 2 inputs and produces one output which is sum of all inputs received. Machine Z takes the outputs of machines X and Y as inputs. Machine Z multiplies its inputs and produces it as output. Based on the given information and the below figure find the missing value ?

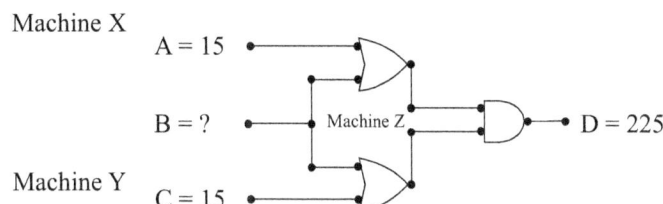

(A) 0 (B) 15

(C) 18 (D) 5

25. Rosy's farm is in the shape of a trapezoid with an area of 21.15 sq.km. If the lengths of the parallel sides of the farm are 7.1 and 2.3 find the distance between them ?

(A) 3.5 (B) 2.25

(C) 5.45 (D) 4.5

26. A cylinder with a height of 8 ft has a volume of 1608.5 cu.ft. Find the diameter of the cylinder ?

(A) 16 ft (B) 8 ft

(C) 4 ft (D) 20 ft

27. The sum of the digits of a two digit number is equal to 11. If the digits of the number are reversed the value is decreased by 63. Find the number ?

(A) 54 (B) 92

(C) 29 (D) 65

28. Find the number of unique permutations of the letters in the word **IDENTIFIED** ?

(A) 220,500 (B) 192,350

(C) 151,200 (D) 84,175

Warm up 14

1. Evaluate $m - \dfrac{mp}{6} + p$; where $m = 5$, and $p = 6$

   (A) 6   (B) 5

   (C) 12   (D) 4

2. Find the solutions of $|9x - 6| = 30$

   (A) $\left\{\dfrac{9}{4}, -4\right\}$ (B) $\left\{4, -\dfrac{8}{3}\right\}$

   (C) 4   (D) $\left\{6, -\dfrac{9}{2}\right\}$

3. Evaluate $2^3 \cdot 64^{3p} = 16^{2p}$, Find the value of "p".

   (A) $\dfrac{3}{5}$ (B) $\dfrac{1}{5}$

   (C) $-\dfrac{3}{10}$ (D) 2

4. Find (x, y) by solving the below equations

   $$11x + y = -7$$
   $$11x + y = -6$$

   (A) Infinite number of solutions   (B) No solution

   (C) $(-8, 11)$ (D) $(-8, -11)$

5. A straight line of slope $-\dfrac{1}{3}$ passing through $A(9, 2)$ and $B(-9, y)$. Find the value of x

   (A) 3   (B) 8

   (C) 4   (D) 1

6. Find the equation of a straight line passing through $N(-3, 4)$ and perpendicular to $y = 3x + 4$

   (A) $y = -\dfrac{1}{3}x + 3$ (B) $y = \dfrac{1}{3}x + 3$

   (C) $y = 3x - 1$ (D) $y = -x + 3$

7. Solve for "n" from the below inequalities

$$16k + 19 \geq 17k + 10 \text{ or } 19 + 10k < 11k + 2$$

(A) k ≥ 3

(B) k ≤ −21

(C) k ≤ 9

(D) k ≤ 9 or k > 17

8. The money used in Alaysia is called the Ginggit. The exchange rate is 4 Ginggits for every $1. Find how many Ginggits you would receive if you exchanged $2 ?

(A) 8 Ginggits

(B) 7 Ginggits

(C) 1 Ginggits

(D) 9 Ginggits

9. What is the selling price of a puzzle that is priced $40.00 and when a discount offered on it is 20% ?

(A) $32.00

(B) $48.00

(C) $44.00

(D) $8.00

Practice test 14

1. Larry sold a Pokemon Game at $40.25 and sold it for a 75% profit.
   Find the cost price of the Pokemon Game.

   (A) $19.55   (B) $23.00

   (C) $17.25   (D) $25.75

2. What is the selling price of a i-phone that is priced $210.00 and with levied tax of 5% on the price ?

   (A) $199.50   (B) $10.50

   (C) $220.50   (D) $252.00

3. John added 8 instead of 47 in his calculations. Find the percentage change in his calculation.

   (A) 83% decrease   (B) 39% decrease

   (C) 487.5% decrease   (D) 86.9% increase

4. Peter took a loan of $21,300 at 2 % interest for 6 years. Find the amount he needs to pay at the end of the loan term.

   (A) $2,556.00   (B) $23,987.26

   (C) $21,726.00   (D) $23,856.00

5. Sophia took a loan of $1,410 at 10% interest compounded annually for 2 years.
   Find the amount she needs to pay at the end of the loan term.

   (A) $1,700.05   (B) $1,686.37

   (C) $1,692.00   (D) $1,706.10

6. Lucy bought 10 dresses for a total of $204. Formal dress cost $26 and casual dress cost $12.
   How many number of casual dress did she buy ?

   (A) 4 casual dress   (B) 3 casual dress

   (C) 2 casual dress   (D) 2 casual dress

7. Find the missing terms in the below sequence

   ..., -31, ___, ___, ___, -51, ...

   (A) -34, -38, -42   (B) -33, -36, -39

   (C) -36, -41, -46   (D) -35, -40, -45

8. Find the missing terms in the below sequence

..., 20, ___, ___, ___, ___, ___, -22, ...

(A) 23, 28, 33, 38, 43   (B) 13, 6, -1, -8, -15

(C) 27, 34, 41, 48, 55   (D) 25, 30, 35, 40, 45

9. Find the missing terms in the below sequence

..., -2, ___, ___, -250, ...

(A) -4, -8   (B) -2, -4

(C) 5, 25   (D) -10, -50

10. Write the explicit formula for the below number series

11, 111, 211, 311, 411, ...

11. If $P(A) = \dfrac{1}{4}$ $P(A \text{ and } B) = \dfrac{1}{8}$ then $P(B) = $?

(A) $\dfrac{3}{10}$ (B) $\dfrac{3}{20}$

(C) $\dfrac{1}{2}$ (D) $\dfrac{11}{20}$

12. If $P(A \text{ and } B) = \dfrac{33}{100}$ $P(A|B) = \dfrac{3}{5}$ then $P(B) = $?

(A) $\dfrac{11}{50}$ (B) $\dfrac{11}{20}$

(C) $\dfrac{7}{100}$ (D) $\dfrac{2}{5}$

13. If $P(B) = \dfrac{11}{20}$ $P(A \text{ and } B) = \dfrac{143}{400}$ then $P(A) = ?$

    (A) $\dfrac{3}{10}$ (B) $\dfrac{9}{50}$

    (C) $\dfrac{13}{50}$ (D) $\dfrac{13}{20}$

14. 12 gal. of a 79% sugar solution was mixed with 4 gal. of a 27% sugar solution. Find the concentration of the new mixture.

    (A) 95 %   (B) 75 %

    (C) 20 %   (D) 66 %

15. A basket contains four apples and eight peaches. Lucy randomly selects a piece of fruit and then return it to the basket. Then she randomly selects another piece of fruit. Find the probability of the first piece of fruit is an apple and the second piece is a peach.

    (A) $\dfrac{2}{9}$ (B) $\dfrac{2}{55}$

    (C) $\dfrac{1}{4}$ (D) $\dfrac{1}{16}$

16. A cooler at a local sports club contains thirteen bottles of sports drink: eight lemon-lime flavored and five orange flavored. Don randomly grabs a bottle and gives it to John. Then, he randomly grabs another bottle for himself after their tennis game. Find the probability of John getting a lemon-lime and Don getting an orange flavoured sports drinks.

    (A) $\dfrac{6}{55}$ (B) $\dfrac{7}{26}$

    (C) $\dfrac{35}{144}$ (D) $\dfrac{10}{39}$

17. The sum of the digits of certain two digit number equals to 10. When you reverse the digits the value increases by 18. Find the number.

    (A) 64   (B) 46

    (C) 55   (D) 37

18. The sum of the two numbers is 184. If one-third of a number exceeds one-seventh by other by eight then find the largest number.

    (A) 112
    (B) 72
    (C) 104
    (D) 80

19. Three solid cubes of sides 1 in , 6 in and 8 in are melted to form a new cube. Find the surface area of the cube so formed.

    (A) 9 sq.in
    (B) 729 sq.in
    (C) 486 sq.in
    (D) 81 sq.in

20. Rak mixes 8 litres of brand A drink with 4 litres of brand B drink which contains 38 % fruit juice. Find the percentage of fruit juice in brand A if the mixture contains 36 % fruit juice.

    (A) 55 %
    (B) 35 %
    (C) 19 %
    (D) 10 %

21. Find the number of unique permutations of the letters in the word **THINKING** .

    (A) 3,800
    (B) 15,240
    (C) 12,960
    (D) 10,080

22. Find the number of unique permutations of the letters in the word **DIVIDEND**.

    (A) 3,360
    (B) 5,240
    (C) 4,320
    (D) 2,560

23. Average scores of 36 students from class A is 40 and 44 students from class B is 35. Find the average scores of both the sections together.

    (A) 37.25
    (B) 34.5
    (C) 28.25
    (D) 40.5

24. The speed of a bus is seven-eighth of the speed of a train. If the train covers 576 mi in 8 hours then how far will the car travel in 5 hours ?

    (A) 342 mi
    (B) 315 mi
    (C) 375 mi
    (D) 393 mi

Quant Q
TJHSST

Practice Test - 14

25. Danny sells a board game at a profit of 12 %. If he bought it for 20 % less amount and sold it for $6 more he would have made a 50 % profit. Find the cost price of the board game.

   (A) $100   (B) $50
   (C) $75   (D) $125

26. Find $f(8,5)$

$$f(x,y) = \begin{cases} f(x-y, y+x) & \text{when } x > y \\ x + y^2 & \text{otherwise} \end{cases}$$

   (A) 172   (B) 207
   (C) 157   (D) 252

27. Find $f(a,b)$ where $a > b$ and $a = b - a$

$$f(x,y) = \begin{cases} f(x+3a, y+b) & \text{when } x < y \\ x + y^2 & \text{otherwise} \end{cases}$$

   (A) $4a + 4a^2$ (B) $8a + 4a^2$
   (C) $8a + 8a^2$ (D) $4a + 8a^2$

28. Neel travels from his office to the mall at a rate of 20 mi/hr and he walks back to the office at 15 mi/hr. If it takes 5 hours 50 mins for a total travel then the distance from the office to the mall is ?

   (A) 25 mi   (B) 35 mi
   (C) 50 mi   (D) 40 mi

Warm up 15

1. Evaluate $-\dfrac{10}{2}(|y| + x)$; where $x = -4$, and $y = 9$

   (A) −21   (B) −20

   (C) −31   (D) −25

2. Find the solutions of $|3n - 8| = 22$

   (A) $\{-7, -5\}$ (B) $\left\{7, -\dfrac{37}{5}\right\}$

   (C) $\left\{2, \dfrac{2}{7}\right\}$ (D) $\left\{10, -\dfrac{14}{3}\right\}$

3. Evaluate $36^{a-2} = \dfrac{1}{216}$, Find the value of "a".

   (A) $-\dfrac{6}{7}$ (B) $\dfrac{1}{2}$

   (C) $\dfrac{3}{7}$ (D) $-\dfrac{13}{7}$

4. Find (x, y) by solving the below equations

   $7x - 3y = 15$
   $7x - 3y = -48$

   (A) $(-6, 2)$ (B) $(2, -6)$

   (C) No solution   (D) $(2, 6)$

5. A straight line of slope 0 passing through $A(6, y)$ and $B(-3, -9)$. Find the value of y

   (A) -9   (B) 1

   (C) -6   (D) 0

6. Find the equation of a straight line passing through $O(-4, 3)$ and perpendicular to $y = \dfrac{2}{3}x + 4$

   (A) $y = -3x - \dfrac{3}{2}$ (B) $y = -\dfrac{3}{2}x - 3$

   (C) $y = 3x - \dfrac{3}{2}$ (D) $y = -\dfrac{1}{2}x - \dfrac{3}{2}$

7. Solve for "x" from the below inequalities

$$11x + 11 < 5 + 8x \text{ or } -16 + 6x > 2x + 12$$

(A) $x < 3$ (B) $x > 7$

(C) $x \leq -15$ (D) $x < -2$ or $x > 7$

8. If the length of square k is doubled, then its area increases by how many times?

(A) 4 (B) 2

(C) 16 (D) 1

9. A box of pencils costs $2.95. A discount of 50% is offered on it today. What is the selling price of the box of pencils?

(A) $3.39 (B) $1.48

(C) $4.43 (D) $2.80

Practice test 15

1. Tony sold a bag of lemons for $3.54 and makes a profit of 20%. Find the cost price of the bag of lemons.

   (A) $1.59   (B) $2.95

   (C) $2.66   (D) $2.36

2. Rak pays $40.39 for a monopoly game. If the store charges 1 % of tax find the price of the monopoly game before the tax.

   (A) $40.40   (B) $33.99

   (C) $39.99   (D) $39.59

3. Rak added 47 instead of 59 in her calculations. Find the percentage change in her calculations.

   (A) 12% increase   (B) 20.3% decrease

   (C) 125.5% decrease (D) 25.5% increase

4. Fiona gave a personal loan to Mary an amount of $400 at 2% interest. If Mary repays it after 2 years, how much amount did Fiona received ?

   (A) $416   (B) $16.16

   (C) $8   (D) $416.16

5. Nora gave a personal loan to Hazel an amount of $16,900 at 14 % interest compounded annually. If Hazel repays it after 3 years, how much amount will Nora receive ?

   (A) $23,998.00   (B) $25,031.68

   (C) $25,038.09   (D) $25,020.46

6. Find the missing terms in the below sequence

   ..., -10, ___, ___, ___, -410, ...

   (A) -112, -212, -312   (B) -214, -315, -416

   (C) -110, -210, -310   (D) -113, -214, -315

7. Find the missing terms in the below sequence

   ..., -8, ___, ___, ___, ___, ___, -62, ...

   (A) -26, -35, -44, -53, -62   (B) -35, -44, -53, -62, -71

   (C) -33, -42, -51, -60, -69   (D) -17, -26, -35, -44, -53

Practice Test - 15

8. Find the missing terms in the below sequence

   ..., -2, ___, -8, ...

   (A) 7   (B) -4

   (C) -2   (D) 1

9. Write the explicit formula for the below number series

   $\frac{1}{2}, \frac{2}{5}, \frac{1}{3}, \frac{2}{7}, \frac{1}{4}, ...$

   [   ]

10. If $P(A) = \frac{9}{20}$ $P(A \text{ and } B) = \frac{9}{40}$ then $P(B) = ?$

    (A) $\frac{27}{80}$ (B) $\frac{1}{4}$

    (C) $\frac{1}{2}$ (D) $\frac{13}{20}$

11. If $P(B) = \frac{7}{20}$ $P(A \text{ and } B) = \frac{7}{40}$ then $P(A) = ?$

    (A) $\frac{77}{200}$ (B) $\frac{7}{10}$

    (C) $\frac{1}{2}$ (D) $\frac{4}{25}$

12. If $P(B) = \frac{1}{5}$ $P(A|B) = \frac{7}{40}$ then $P(A \text{ and } B) = ?$

    (A) $\frac{9}{100}$ (B) $\frac{3}{10}$

    (C) $\frac{33}{100}$ (D) $\frac{7}{200}$

Quant Q TJHSST

Practice Test - 15

13. Find the sum of the series for the below

$$a_1 = 28, a_9 = 512, r = 2$$

(A) 1014 (B) 1022
(C) 842 (D) 954

14. Find the number of terms in the below series.

$$a_1 = -4, r = 6, s_n = -223948$$

(A) 5 (B) 7
(C) 6 (D) 9

15. Riks magic bag contains eight red marbles and eight blue marbles. He randomly picks a marble and then returns it to the bag before picking another marble. Find the probability of him picking the first marble as red and the second marble as blue.

(A) $\dfrac{1}{8}$ (B) $\dfrac{1}{16}$
(C) $\dfrac{1}{4}$ (D) $\dfrac{15}{91}$

16. Rak flips a coin 7 times. Find the probability of the coin landing heads up every time.

(A) $\dfrac{1}{55}$ (B) $\dfrac{1}{16}$
(C) $\dfrac{1}{128}$ (D) $\dfrac{5}{136}$

17. Find the unique permutations of the letters in the word **FRACTIONS**?

(A) 362,880 (B) 40,320
(C) 45,630 (D) 60,480

18. The difference of two numbers is 11. One-fifth of their sum equals to nine. Find the numbers.

    (A) 28 and 15   (B) 15 and 26

    (C) 33 and 21   (D) 28 and 17

19. Average of four consecutive numbers is 28.5. Find the largest number.

    (A) 30   (B) 24

    (C) 26   (D) 28

20. If the edge of a cube is increased by 50 % find the percentage increase in its surface area.

    (A) 100 %   (B) 25 %

    (C) 125 %   (D) 200 %

21. The capacity of a cylindrical tank is 1848 cu.m with a diameter of 14 m. Find the depth of the tank.

    (A) 12 m   (B) 8 m

    (C) 18 m   (D) 11 m

22. A cylindrical hole with a diameter of 4 inches is cut through a cube the edge of the cube is 5 inches. Find the value of the hallow cylinder in terms of Π.

    (A) 125 -80 Π (B) 125 -20 Π

    (C) 80 Π - 125   (D) 20 Π - 125

23. Two machines X and Y takes 2 inputs and produces one output which is greatest of all.
    Machine Z takes the outputs of machines X and Y as inputs. Machine Z multiplies its inputs and produces it as output. Based on the given information and the below figure find the missing value ?

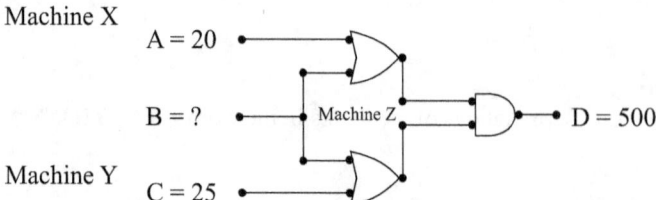

    (A) 29   (B) 25

    (C) 50   (D) 15

Quant Q TJHSST

Practice Test - 15

24. Rak mixes 1 pound brand A peanuts which costs $3 per pound with brand B peanuts which costs $8 per pound. The mixture costs $7 per pound. How many pounds brand B peanuts did she add ?

    (A) 5 pounds (B) 4 pounds
    (C) 3 pounds (D) 7 pounds

25. Sixteen men can complete a piece of work in 24 days. 20 women can complete the same work in 36 days. In how many days will 12 men and 15 women together can complete the same amount of work ?

    (A) $20\frac{3}{8}$ days (B) $18\frac{2}{3}$ days
    (C) $21\frac{5}{6}$ days (D) $19\frac{1}{5}$ days

26. Rik covers x miles in 1 hour 18 mins. He travels three-fifth of the distance at 24 mi/hr and remaining distance by 10 mi/hr. Find the total distance travelled ?

    (A) 16 mi (B) 20 mi
    (C) 25 mi (D) 35 mi

27. The ratio of the radius and height of a right circular cone is 3 : 4. If the curved surface area of the cone is $375\ \Pi\ m^3$. Find the total surface area of the cone.

    (A) $500\ \Pi\ m^3$ (B) $600\ \Pi\ m^3$
    (C) $400\ \Pi\ m^3$ (D) $300\ \Pi\ m^3$

28. Find the number of unique permutations of the word **IDENTIFIED** ?

    (A) 151,200 (B) 149,500
    (C) 154,620 (D) 124,860

Quant Q TJHSST

Answer Keys

#1 Prime Factors
1. A
2. C
3. D
4. B
5. D
6. D
7. B

#2 Numerical Expressions
1. B
2. B
3. C
4. A
5. C
6. A
7. B

#3 Simplify Expressions
1. B
2. D
3. A
4. D
5. A
6. D
7. A

#4 Evaluate Expressions
1. A
2. B
3. D
4. D
5. A
6. D
7. B

#5 Exponential Expressions
1. B
2. B
3. B
4. C
5. D
6. D
7. C

#6 Distance Between 2 Points
1. D
2. D
3. C
4. C
5. B
6. D
7. B
8. C

Quant Q TJHSST

Answer Keys

#7 Midpoint	#8 Equation of a Straight line	#9 Slope (2 points)
1. D	1. C	1. A
2. A	2. C	2. D
3. C	3. B	3. D
4. D	4. D	4. B
5. A	5. A	5. C
6. B	6. B	6. C
7. A	7. C	7. A
	8. C	8. B
		9. A

#10 Slope intercept form	#11 Graph Slope	#12 Find the slope
1. D	1. B	1. C
2. B	2. A	2. A
3. A	3. A	3. B
4. A	4. B	4. B
5. A	5. B	5. D
6. A	6. D	6. B
7. D	7. A	7. B
8. A		

©All rights reserved-Math-Knots LLC., VA-USA
For more practice visit www.a4ace.com
www.math-knots.com

Quant Q TJHSST

Answer Keys

#13 Parallel line slope
1. D
2. A
3. B
4. A
5. C
6. B
7. C

#14 Perpendicular line slope
1. A
2. A
3. A
4. D
5. A
6. D
7. B

#15 Radicals 1
1. B
2. D
3. C
4. B
5. B
6. C
7. C
8. C

#16 Radicals 2
1. A
2. A
3. B
4. C
5. A
6. A
7. B
8. C

#17 Inequalities
1. B
2. D
3. A
4. D
5. C
6. C

#18 One step word problems
1. B
2. B
3. C
4. B
5. A
6. A
7. D
8. A

Quant Q TJHSST

Answer Keys

#19 Circle area
1. B
2. A
3. A
4. D
5. A
6. D
7. C

#20 Volume of a Sphere
1. A
2. C
3. D
4. B
5. C
6. D
7. A

#21 Volume of rectangle, square, prisms
1. D
2. C
3. A
4. C
5. D
6. D

#22 Volume of cone cylinder
1. B
2. C
3. A
4. C
5. B
6. A
7. A

#23 Missing angle 1
1. C
2. A
3. B
4. B
5. B
6. B
7. D

#24 Missing angle 2
1. A
2. C
3. A
4. B
5. C
6. B
7. C

#25 Reflection
1. B
2. C
3. B
4. A
5. A
6. D
7. A

#26 Rotation
1. D
2. B
3. C
4. D
5. A
6. A
7. B

#27 Translation
1. C
2. D
3. B
4. B
5. C
6. A
7. D

Quant Q TJHSST

Answer Keys

Warm up 8

1. D
2. B
3. D
4. A
5. B
6. B
7. B
8. A
9. C

Practice test 8

1. A
2. D
3. C
4. D
5. C
6. A
7. C
8. D
9. A
10. A
11. B
12. B
13. D
14. A
15. D

16. C
17. A
18. A
19. A
20. A
21. C
22. A
23. B
24. C
25. B
26. $\dfrac{-15}{n+3}$
27. B
28. A

Warm up 9

1. B
2. A
3. A
4. C
5. C
6. C
7. C
8. A
9. D

Practice test 9

1. C
2. C
3. A
4. B
5. D
6. A
7. A
8. C
9. D
10. D
11. A
12. C
13. D
14. C
15. B
16. D
17. D
18. B
19. A
20. A
21. A
22. C
23. A
24. D
25. C
26. B
27. D
28. $a_n = -10 + 20n$

Warm up 10

1. A
2. D
3. A
4. D
5. C
6. A
7. A
8. A
9. A

Practice test 10

1. D
2. B
3. A
4. C
5. A
6. A
7. A
8. C
9. D
10. $a_n = -10 + 4n$
11. A
12. D
13. D
14. D
15. B
16. D
17. A
18. D
19. A
20. C
21. C
22. D
23. B
24. C
25. A
26. A
27. A
28. D

Quant Q TJHSST

Warm up 11

1. D
2. D
3. A
4. D
5. C
6. A
7. C
8. D
9. A

Practice test 11

1. D
2. A
3. B
4. D
5. D
6. A
7. C
8. B
9. $a_n = 12.4 + 1.3n$
10. B
11. D
12. C
13. D
14. A
15. D
16. C
17. A
18. A
19. B
20. A
21. C
22. C
23. B
24. A
25. C
26. A
27. D
28. B

Quant Q
TJHSST

Answer Keys

Warm up 12

1. D
2. B
3. A
4. A
5. C
6. D
7. C
8. B
9. D

Practice test 12

1. A
2. D
3. C
4. A
5. B
6. A
7. C
8. D
9. C
10. $a_n = -232 + 200n$
11. D
12. C
13. B
14. A
15. D
16. D
17. D
18. A
19. A
20. B
21. A
22. D
23. A
24. A
25. B
26. D
27. A
28. C

Quant Q TJHSST

Answer Keys

Warm up 13

1. A
2. C
3. D
4. D
5. B
6. D
7. A
8. C
9. A

Practice test 13

1. A
2. C
3. A
4. A
5. A
6. D
7. A
8. D
9. D
10. $a_n = \dfrac{n}{2^n}$
11. D
12. C
13. C
14. A
15. D
16. A
17. B
18. C
19. D
20. A
21. A
22. A
23. C
24. A
25. D
26. A
27. B
28. C

Warm up 14

1. A
2. B
3. C
4. B
5. B
6. A
7. D
8. A
9. A

Practice test 14

1. B
2. C
3. A
4. A
5. D
6. A
7. C
8. B
9. D
10. $a_n = -89 + 100n$
11. C
12. B
13. D
14. D
15. A
16. D
17. B
18. B
19. C
20. B
21. D
22. A
23. A
24. B
25. C
26. A
27. D
28. C

Quant Q TJHSST

Answer Keys

Warm up 15

1. D
2. D
3. B
4. C
5. A
6. B
7. D
8. A
9. B

Practice test 15

1. B
2. C
3. B
4. A
5. C
6. C
7. D
8. B
9. $a_n = \dfrac{2}{n+3}$
10. C
11. C
12. D
13. B
14. B
15. C
16. C
17. A
18. D
19. A
20. C
21. A
22. B
23. D
24. B
25. D
26. B
27. B
28. A

www.ingramcontent.com/pod-product-compliance
Lightning Source LLC
Chambersburg PA
CBHW080739300426
44114CB00019B/2627